This book is to be returned on or before
the last date stamped below.

Sharecroppers of the Sertão

Allen W. Johnson

Sharecroppers of the Sertão

*Economics and Dependence on a
Brazilian Plantation*

1971
Stanford University Press
Stanford, California

10394

Stanford University Press
Stanford, California
© 1971 by the Board of Trustees of the
Leland Stanford Junior University
Printed in the United States of America
ISBN 0-8047-0758-8
LC 74-130827

For David

Acknowledgments

In beginning my fieldwork, I received substantial help from the Institute of Anthropology of the University of Ceará and its director Sr. Fontenelle, and from Srs. João Pompeio and Eleo Gois. Professor Bernard J. Siegel sponsored my research and contributed practical aid and insights during all phases of the research. The Agricultural Development Council and the National Science Foundation provided the material support for the study. Professors George Dalton, Daniel Gross, Marvin Harris, Andrew Vayda, and Charles Wagley read the manuscript and offered many helpful comments and criticisms. To all of these I am most grateful. I wish also to express my deepest thanks to Sr. Clovis Holanda, owner of Boa Ventura, for his understanding and warm support during my stay on his *fazenda*. Finally, I would like to thank Sr. Zeca Paiva, informant and friend.

A.W.J.

Contents

Tables and Figures

Eight pages of illustrations follow page 68

Sharecroppers of the Sertão

One

Introduction

The following pages contain an ethnographic account of the economic behavior and concerns of the resident workers on a plantation in Ceará, Brazil. In recent years, there has been a growing interest in the social and economic structure of plantations, *haciendas*, *fincas*, and other large-scale agricultural enterprises characteristic of tropical areas (e.g. Geertz 1963, Harris 1964, Hutchinson 1957, Miller 1967, PAUSSM 1959, Steward *et al.* 1956, Wagley 1957, Wolf and Mintz 1957); and many of the problems dealt with here have been raised in earlier studies. However, it is hoped that this study will prove to be a unique contribution within the general area mapped out by previous students.

Throughout the work, I have emphasized two points that have been stressed much less in other works. First, my methodological concentration has been on measurement and specification. For example, the exact numbers of man-days of labor involved in all phases of the agricultural cycle are given, and are later compared with the exact amounts of goods produced and consumed. My hope is that by avoiding nonspecific statements whenever possible I have enabled the data presented here to be more fruitfully compared with data in the hands of other researchers.

A second major emphasis of this work is on the search for security, which underlies much of the plantation workers' eco-

nomic behavior, from their planting behavior to their relations
with fellow workers, or members of the upper class. To understand this search fully, it was necessary not merely to observe
the workers, but also to ask at all times how the worker himself perceived the context within which he was aiming to increase his security. It is generally true that many of the conclusions drawn here would not have been possible without a
constant comparison of the workers' attitudes with field observations collected at the same time.

THE FAZENDA

The Fazenda Boa Ventura, the subject of this study, is in the
sertão, the semi-arid backland of northeastern Brazil. In this
part of the world, a *fazenda* may be practically any landholding of more than a few hectares in size, regardless of its organization and operation. It has become common to regard the
traditional hacienda and the more modern plantation as two
basic kinds of rural social systems in Latin America (Wolf and
Mintz 1957: 380). Further, "plantations and haciendas should
probably be thought of as the polar extremes of a taxonomic
continuum" (Harris 1964: 45). The term fazenda, as used in
northeastern Brazil, seems to range over the whole continuum.
Wolf and Mintz enumerate the general characteristics of the
continuum as follows: employment of many workers on large
land areas to produce a culturally defined surplus that is sold
on a market; the transfer of some of the laborers' produce to a
ruling class (that is, there is class stratification); specialization
in one crop to supply market needs; accumulation of capital;
and finally, a political and legal system that supports the acquisition of a stable labor supply and the transfer of surpluses
from producers to owners. Fazenda Boa Ventura shares all these
characteristics, and hence belongs on the continuum. Although
its size (about 2500 hectares and 50 families) is above average
for the area, it is closer to the hacienda than to the plantation
extreme.

Boa Ventura is close to the hacienda pattern in deriving its limited capital from one family, rather than from a corporation. Like the hacienda, it can obtain labor easily because no alternatives exist for the laborers (most workers are landless and cannot acquire land). In plantation areas, by contrast, workers were or are bound by slavery, debt peonage, or labor laws to work for a *fazendeiro* (landlord). As on haciendas, subsistence plots are available for the workers, whereas plantations pay wages exclusively. However, like the plantation type, Boa Ventura produces a cash crop (cotton) for the world market, and raises staple crops and cattle for sale in the local markets. Plantations and haciendas, as ideal types, also differ from one another in the absence or presence of close personalistic ties between worker and landowner. Here, Boa Ventura is in flux: the man who owned the fazenda until five years before this study allowed personal ties between himself and most of the workers, whereas the present owner severely limits them. Indeed, in most respects the Fazenda Boa Ventura could easily pass from hacienda into plantation simply by changing owners. For example, it is now rumored that the administrators of the highly capitalized, corporation-owned neighboring plantation want to buy Boa Ventura and transform it into a banana and cotton plantation worked by wage laborers under capital-intensive methods.

Almost all the labor requirements of Boa Ventura are met by the resident workers (*moradores*), who live in houses provided by the fazenda throughout the year. There are *meieros*, who give shares of their produce, and *diaristas*, who give specified amounts of labor to the fazenda in return for various rights, including rights to land and a house. All workers are one or the other of these except the cowboy (*vaqueiro*), who has a status and contract of his own.

It is difficult to say whether the moradores of Boa Ventura should be regarded as peasants. Wolf (1955: 453) emphasized that the agricultural worker who does not retain "effective control of the land" (i.e. a "tenant") should not be considered a peasant; later (1956), he opposed "agricultural workers" to "peasants." By this definition, the workers of Boa Ventura are not peasants, since their control over the land is subject to the final authority of the landlord. However, in many other respects the moradores of Boa Ventura do seem to be part of the world's peasantry. Wolf's other criterion, that the peasant is an agricultural producer aiming at subsistence rather than reinvestment, clearly applies to Boa Ventura's resident labor force (Wolf 1955: 453–54).

Passing from economic to social criteria, we find complete agreement between conditions on Boa Ventura and Fitchen's characterization of the peasantry as a social type (Fitchen 1961: 114–19): kinship is relatively unimportant; the household, within which property is held in common, is the basic unit; the effective community is defined territorially, not genealogically; sharp horizontal (i.e. class) distinctions are present, and sharp vertical (i.e. lineage) distinctions are not; finally, a larger culture exists outside the peasantry, and the peasant is aware of it. Wolf himself has recently changed his definition to emphasize the political and economic relation of the agricultural workers to the dominant upper class that appropriates a share of the workers' produce (Wolf 1966: 3–4). Certainly, the moradores of Boa Ventura fit this usage. In any case, the residents of Boa Ventura are here considered as peasants—peasants as much like poor landowning peasants in Mexico, for example (Foster 1961), as they are like other plantation or hacienda workers in South America.

For the most part, I have considered Boa Ventura as a unit, and the assertions made and figures computed apply to its entire population. This is in keeping with the fact that it is an enterprise ultimately controlled by one man and clearly set apart from other enterprises in the area. But this kind of treatment has an artificial aspect to it, since the formal boundaries of the fazenda do not coincide with its residents' isolation from outside ties and contacts. The landlord has many important outside contacts with traders, other fazendeiros, and financiers. The workers belong to a local social system that crosses fazenda boundaries. Most of them have relatives on nearby plantations with whom they carry on continuing exchanges. Many of them are known by neighboring fazendeiros or their managers, and at times work for them as day laborers. The majority have contracted debts in shops outside the fazenda. And, in general, they are aware of outside possibilities for themselves: as moradores on other fazendas; as employees in nearby towns; or even, in a few cases, as small landholders in their own right. At the time of this study, all the workers, with one exception, had been born elsewhere; and most had lived on several fazendas before coming to Boa Ventura.

In these few pages, I have already found occasion to mention social classes several times. This is as it should be, for the presence of class distinctions is a basic fact of life for the morador of Boa Ventura; and it seems to obscure the differences that exist between the moradores themselves. Even though marked wealth differences exist between the moradores, they do not verbalize these differences, but continue to think of themselves as essentially equal, different only from the higher classes with whom they come into contact.* Possessing pride and integrity in their dealings with each other, they are aware of their own weakness and vulnerability with respect to the landlord, and consequently seek a personalized relationship with him as a defense against an uncertain physical and social environment. This feature is characteristic of plantations, particularly those √of a traditional type.† "Paternalism stands out as one of the major factors in the complicated organizational hierarchy of plantation life. . . . At its highest level it is a form of unwritten social security which works well." (Hutchinson 1957: 71). This subject will be developed in Chapter Seven.

AIMS OF THE BOOK

This book has grown directly out of my fieldwork, and a brief account of that work will serve to introduce the book. The field-

---

* Contrast this with the real importance of differences in wealth between Mexican peasants, as reported by Cancian (1965).
† See Hutchinson (1957: 179–84) and Mintz (1953: 250) for discussions of the modern trend toward a breakdown of paternalism.

work was conducted in the interior of Ceará, Brazil, from August 1966 to August 1967. During this period, my wife and I spent one month on each of two plantations, conducting preliminary research and attempting to locate a large, technologically advancing plantation on which to concentrate. The remaining ten months were spent on Fazenda Boa Ventura, which came closer to the desired type than any of several other possibilities.

The fieldwork that yielded the substance of this book was undertaken to gain an understanding of the total economic life of the resident laborers on a fazenda, regarding this life as a form of ecological adaptation. The fazenda was considered an economic unit occupying a particular "ecological niche," as that term is used by cultural ecologists (Barth 1956); this led to an inquiry based on three broad questions. First, which aspects of the environment are relevant? That is, which ecological conditions are seen as important by the workers, and how do they conceptualize them? And what other conditions seem to be important, regardless of the workers' awareness of them? Second, what is the nature of the labor force? What are the qualities and capacities of the worker as an agricultural laborer—his technology, his motivation, and his ability? What range of economic specializations is represented? Third, what are the social implications of applying this labor force to the problems presented by the ecology (i.e. the "social relations of production")?

The publication of Steward's *Theory of Culture Change* (1955) did much to introduce the concept of cultural ecology into anthropology, and many later works share Steward's basic orientation (e.g. Goldschmidt 1965, Leeds 1961, Sahlins 1962). Much of this work has been focused on evolutionary schemes of human social history, but still more has been done in applying cultural ecology to specific cases (in addition to the above citations, see Leeds and Vayda 1965, Steward *et al.* 1956).

Basic to the ecological approach is the notion that survival in a particular environment is the fundamental problem faced by man; and that the behavior aimed at solving this problem is the most important behavior for us to study if we wish to understand cultural process and change. This has led to the concept of a culture "core" made up of "those features most closely related to subsistence activities" (Steward 1955: 37; see also Leeds 1961: 6, and Sahlins's contrast between "fundamental" and "superstructural" cultural elements, 1962: 28). This study, then, is linked to the work of Steward and others, for it attempts to describe the way of life of the Boa Ventura workers as a direct result of the process of adaptation to the ecological setting in which they function.

For the most part, the students of human ecology discussed above are interested in problems of culture change. Yet here I have concentrated on the present conditions found on Boa Ventura, only secondarily raising the question of how they came to be and practically ignoring the question of where the fazenda may be going. In my opinion, an understanding of on-the-ground conditions must obviously precede a search for areas where cultural changes are likely to take place; I view this work as a preliminary step, and hope that my decision to organize it as a search for relations between environment and human behavior will make questions about change much easier to ask.

Plantations are not yet well-known economic systems, and objective descriptions of them are sorely needed; that is the main justification for this book. I believe that my data have numerous implications for the understanding of swidden agricultural systems, economic development, class-structured agrarian societies, patron-client relations, dyadic contracts, migration, and many other problems of social anthropology. In many cases, I have attempted to outline these implications for the reader; but in other cases I have thought it best to leave theorizing to those more expert in a particular area.

I collected the data reported on here without using assistants or translators. Early in the fieldwork, I reached an important conclusion: I could trust my informants to give me information that was both honest and accurate in regard to their economic behavior, in part because of the general openness of rural Brazilians, and in part because of the great attention they paid to their own subsistence activities. Workers frequently measured their fields; and since my own measurements were almost identical to theirs, I decided to accept their estimates rather than measure all the fields myself. In the same way, workers always measured their crop harvests, storing them in liter jars or 20-liter cans in their houses. Since I had to leave Boa Ventura before the entire harvest was collected for 1967, I recorded harvests from previous years by asking informants how much they had harvested; they always answered without hesitation, and I believe that the information was reasonably accurate.

The decision to trust my informants was admittedly an expedient. However, being relieved of time-consuming direct measurements, I could, through systematic questionnaires, elicit a larger quantity of information from a larger number of individuals than if I had insisted on direct observation. Of course, there is an indeterminate loss of accuracy in this method; and those readers who have worked in areas where informants habitually lie and distort, or where they are careless in keeping track of their subsistence activities, will be somewhat skeptical of the data reported herein. I should point out, however, that for an hour spent closeted with an informant I usually spent another hour collecting information directly through observation. The systematic questionnaire data are here presented primarily in tables, whereas the other data serve to flesh out the text.

The organization of the book is based on the problem of adaptation. The next chapter is concerned with the material environment, and Chapter Three deals with the circulation and settlement of the human element in this system. Chapters Four and Five concentrate on the subsistence techniques by which the people produce necessary goods and services. Finally, the sixth and seventh chapters explore the social relations that are implied by or seem to grow out of subsistence activities.

At all stages of this kind of inquiry, one encounters the problem of what to consider and what to ignore. By and large, the activities associated with food production and the provision of other material necessities should quite clearly be included. But greater difficulty is encountered at the social level. There is no a priori way of saying which aspects of the social system are, in Steward's words, "most closely related to subsistence activities." Just what does this phrase actually mean, and where does one draw the line between core features and superstructural features? Steward points out that this is an empirical problem, to be determined anew for each different social system: "Although technology and environment prescribe that certain things must be done in certain ways if they are to be done at all, the extent to which these activities are functionally tied to other aspects of culture is a purely empirical problem" (Steward 1955: 41).

For the Fazenda Boa Ventura, I have tried to specify the social relations that seem to be directly related to the problem of survival faced by the moradores. It is important to note that many social relations are not entirely inside or outside the culture core. Social relations that have an adaptive aspect may also have attributes that seem to contribute little to an individual's survival or to the well-being of the group as a whole. What is especially remarkable about Boa Ventura is that so much of the social life of its people is closely related to problems of survival. The social relations discussed here—between kinsmen, neighbors, and patrons and clients—are almost always undertaken with the aim of sustaining one's material security (although at a relatively marginal level), and are conceptualized as such by the workers.

Informed as it is by the ecological considerations, this work remains basically an ethnography; and even though much ethnography has been stimulated by an ecological approach in the past, it has not always been the best ethnography, for two reasons.

First, not enough attention has been paid to getting an adequate picture of how the people being studied think about their environment and the challenge of surviving in it. This "emic" orientation has been left for the most part to the ethno-scientists (e.g. Frake 1962)—whose work has often been considered limited, with some justification, because it has avoided the broad questions of behavioral adaptation and change in response to specific material environments (Berreman 1966, Harris 1968: 591–92). But we should recognize that ethnoscience has done much to develop more reliable techniques for eliciting and reporting an "informant's-eye view"; and these techniques are at our disposal.

At present, it appears that the overriding concern of the non-ethnoscientific ecological studies cited earlier has been to show how cultures, as bodies of traditions, have adapted to particular environments and how this adaptation has dictated the behavior of the individual carriers of the culture. There is a strong tendency to ignore the reasoning of informants, and to assign meaning to their behavior according to the categories of the observer's theory rather than native categories. But it is clear that both approaches are needed. On Boa Ventura, to be sure, there are some patterns of behavior of which the workers themselves are not aware. For instance, in migrating, workers seldom return to a plantation on which they have once lived; yet they are not aware of any value this may have, or even aware that they are doing it. But it would be misleading to assume that there is no reasoning behind much of what the moradores do. In subsistence activities, one can see wide differences in individual behavior, and it is obvious that individuals reason to different conclusions because of their different personal capacities. At times, different decisions are reached because two moradores disagree fundamentally about the parameters of the problem: they simply do not share the same "cultural matrix" (cf. Wolf 1956: 175).

The information from Boa Ventura strongly suggests a closer look at the notion, often implicit in anthropological studies, that subsistence behavior is controlled entirely by tradition. On Boa Ventura, I found several moradores who were conducting experiments with new methods and crops. In some cases, part of a crop would be planted in the new way, and a section planted traditionally acted as a control. This is by no means an isolated case. Among the Hanunóo of the Philippines, Conklin (1957: 110) found experimental gardens beside many houses, in which the natives tested new plant types or techniques before using them in the main swidden plots. Manners (1956: 110) reports experimentation among Puerto Rican coffee growers. Even though traditions, especially in the subsistence sphere, may represent an adaptation to environmental conditions, one should not assume that there is no variety in behavior; in fact, members of a society may often disagree, even on basic concepts. In evolutionary terms, this variation is necessary to provide the "mutations" from which new forms of behavior are selected.

The second major deficiency often found in ethnographies written by anthropologists of the "adaptivist" persuasion is a lack of specificity or quantification.* Steward (1955) emphasizes that the search for "parallels" between superficially dissimilar societies is essential to his theory. But it is obvious that the reliability of such parallels depends on the reliability of the ethnographic work from which the facts are drawn. In speaking of swidden (slash-and-burn) technology, or in comparing plantation economies, it is not enough to know that manioc or maize is planted; the amount of seed and the area and quality of land planted are also important. Likewise, in comparing aspects of production and consumption, one must establish concrete and comparable measures of labor productivity and household consumption. In describing economic relations, one should determine the kinds and quantities of goods exchanged in each sort of transaction: for example, on Boa Ventura different goods are exchanged in different quantities between a worker and another worker, between a worker and a shopkeeper, and between a worker and a landlord. Generally, ethnographers outside of the adaptivist circle seem to have been

* The outstanding recent exception to this charge is Rappaport's study of New Guinea (1967).

more aware of this need for specificity (Conklin 1957, Salisbury 1962, Tax 1953).

A noted economist has asserted that the economy of agrarian society remains "a reality without a theory" (Georgescu-Roegen 1964). To the extent that this is true (particularly of peasant society, which is what he was really referring to), I believe that it results from a lack of proper data from which to draw comparisons and hypotheses.

Two

The Environment

In many respects, an environment—physical, social, or cultural—limits the activities of its inhabitants and determines the specific lines their activities take. We must now describe the broad features of the environment within which the moradores of Fazenda Boa Ventura exist. Later, it will be possible to connect specific environmental characteristics with specific activities and beliefs of the moradores.

The area in which this study was done is usually called the "Northeast" of Brazil. In reality, there are two Northeasts. What one usually thinks of as Brazil's Northeast is the humid coastal region of cacao and sugar production. It is there that Brazil's colonization began, and it is there that recent political events (especially the formation of the Peasants' Leagues) have attracted international attention. But immediately behind the coast lies the broad sertão, or backland, a semi-arid region covered with an unattractive scrub brush called *caatinga*. Boa Ventura lies in the middle of this dry, inhospitable region.

Fazenda Boa Ventura is situated in the center of the state of Ceará, and is accessible by an uncompleted national highway running southwest 180 kilometers (112 miles) from the capital city of Fortaleza on the coast (see Figure 1). To be more precise, it lies 39° 30′ W and 4° 55′ S, in the northerly portion of Brazil's "drought polygon" (an area embracing parts of Ceará

Figure 1. The state of Ceará, showing the location and physical environment of Fazenda Boa Ventura.

and several other northeastern states). A traveler can reach the fazenda by bus and jeep in five hours during the dry season; but in the wet season it takes eight or nine hours to make the same trip.

Ceará can be divided into three climatic zones: the *litoral*, a low, moist strip along the coast; the *serra*, or rocky mountains (up to about 1,000 meters in elevation), which occur sporadically throughout the state; and the semi-arid sertão. The first two regions are characterized by a reliable rainfall sufficient for intensive agriculture, whereas the third is not.

Boa Ventura is situated in typical sertão conditions, and receives an average of 28 inches of rainfall a year. But this figure, taken by itself, is very misleading, for the fundamental fact about rainfall here is its variability. As the residents say, "If it isn't eight, it's 80; if there's not a drought, there's a flood." The most dramatic variation is from year to year. For example, a spot that received 40 inches of rain in a wet year could receive less than eight inches in a dry year. In addition to this annual variation, geographical and seasonal variations must be considered. The occurrence of a heavy rainfall in the sertão as a whole does not guarantee that any particular place will receive enough water. Even in the most ideal years (during one of which this study was done), some local areas will suffer from too much or too little rain.

The sertão experiences two sharply defined seasons. This is just about the only certainty in its unpredictable climate. Winter, the rainy season, extends from January through July, and summer, the dry season, from August through December. But even the seasons are unpredictable. Winter may start as early as December or as late as March. There may be a false start followed by a long dry stretch that will kill any seedlings planted. Rain, though unlikely, is still possible in every month of the year, and may be harmful if it falls in the summer months. It is clear that the farmers of the sertão find no security in the weather; they may be enriched or virtually impoverished by it; but however much they study it, in the end neither they nor the scientists in town can say what the winter will bring. As will be shown, these farmers, in response to such a variable climate, are very good at acting on many possibilities rather than one or two strong probabilities.

Boa Ventura is only a few degrees below the equator, and like other equatorial regions it has a fairly stable temperature. Average daily temperatures range from +24 to +32 degrees centigrade, with a very small seasonal variation; the extremes recorded in the sertão are a low of +17 degrees and a high of +37 degrees. Because the humidity is seldom excessive and because there are steady breezes throughout the year, any place out of the sun is very comfortable. Nonetheless, the intense sunlight generally proscribes noonday labor; and except for the seasons when the work load is heaviest, the laborers are usually inactive between 11 A.M. and 1 P.M.

In the sertão one finds no high mountains, broad alluvial valleys, or wide plains; rather, the countryside resembles "a shapeless heap of mountainous ruins" (Da Cunha 1944: 17). One occasionally encounters actual mountains, but the usual landscape is a rough scrubland sprinkled with rock outcroppings of moderate size. Most farmers, in order to survive, plant hillsides in addition to whatever flat pieces they may acquire. The best lands are found along the river beds, and these areas are always cleared and kept under continuous cultivation. The hilly uplands, where most farming takes place, are not nearly as fertile, giving good yields the first year but dropping rapidly thereafter; hence the "bush fallow" cultivation system, in which fields are farmed for two years and left fallow for about eight.

In a sense, there are two natural plant covers in the sertão: a xerophytic cover that is dominant during the dry season, and a hydrophytic cover dominant during the rains. In the dry season, one sees only the dry remnants of the wet-season plants, along with the cacti and thorny growth characteristic of desert environments. But after the arrival of the winter rains, "the great tempests that put out the silent conflagration of the drought," the change is sudden and complete. During the drought, the caatinga has been almost invisible, lying dormant in thick roots beneath the surface and beyond the effects of the sun, or in long-lasting seeds. Now, the backlands suddenly turn green as the short, tough trees and scrub brush leaf out, and the grasses sprout. In a matter of days, the sertão is hardly distinguishable from the humid littoral. Under the usual farming methods, large sections of the hilly landscape will be left in caatinga at any one time. From this growth, the farmer takes wood for his houses, fences, tool handles, and fires. In winter, it serves as a pasture for his cattle, which in turn fertilize the land; and as long as it is permitted to grow untouched, the caatinga itself enriches the earth, restoring it to usable fertility.

Owing perhaps to the long habitation of this area by man, wildlife for hunting is rare; squirrels and small birds are the usual game. In general, as far as wildlife is concerned, the environment offers more a threat than a boon. Rattlesnakes, coral snakes, and fer-de-lance are quite common. Although they are not as deadly, tarantulas and scorpions are common in the fields, and may even crawl into hammocks at night.

THE SOCIOCULTURAL ENVIRONMENT

Little is really known about the native population of Ceará before Portuguese colonization. The first reports, from the years following 1600, record the presence of Tupinambá Indians; but it may be that the Tupinambá were pushed into the area by the earlier white occupation of Paraíba and Rio Grande do Norte (Girão 1947: 35). In any case, a non-Tupi people, the Teremembé, was also present at the time of the Portuguese conquest, and was probably a remnant of Ceará's earlier inhabitants. Both this tribe and the Tupinambá were eradicated by the Portuguese, and most of their culture is lost, except for a few aspects of Tupinambá technology. However, this technology has not only persisted but has continued to dominate the subsistence sector of the rural economy of Ceará.

Whereas the Teremembé were hunters and gatherers (Metraux 1948a: 573), the Tupinambá were tropical forest agriculturalists, subsisting chiefly on manioc and maize. In addition, they planted sweet potatoes, lima and kidney beans, pumpkins, peanuts, pineapple, pepper, gourds, tobacco, and cotton. "The Tupinambá cleared farmland in the forests near their villages, felling the trees with stone axes and burning them a few months later. The ashes served as fertilizer. Women did all the planting and harvesting. At the beginning of the dry season they set out manioc cuttings and sliced tubers, and planted maize and beans in holes made with pointed sticks. They did no other work except some occasional weeding. . . . To increase the cotton yield, they thinned the trees twice a year." (Metraux 1948b: 99.) With certain important but limited changes, the present population of the sertão employs the same crops and methods that their Indian predecessors used, particularly in subsistence agriculture. Many Portuguese introductions, such as bananas, sugar cane, and cattle, have merely been superimposed on the native economy as commercial crops for an urban or world market.

The main force of the Portuguese conquest in Northeast Brazil hit the humid coast to the south of Ceará. Ceará itself was too dry for commercial agriculture, and was therefore of little interest to the Portuguese proprietors, except insofar as it was necessary to use military force in the region to forestall colonization by the French or to round up Indians for plantation labor on the coast. The lands to the south were ideal for the planting of sugar cane, a highly profitable cash crop; and it was there that the Portuguese occupied themselves throughout the sixteenth and seventeenth centuries.

On the sugar coast, little land remained free for the production of foods. To support the cane workers and their families, the barren sertão was gradually opened up to cattle raising (Prado Junior 1963: 43). At first, cattle raising was limited to the immediate hinterland of the sugar-growing regions in Bahia and Pernambuco. But through time, and particularly toward the end of the seventeenth century, cattle raising spread southward into Minas Gerais and northward, through Ceará, into

Maranhão. The two major cattle routes from North to South both traversed Ceará, one through Fortaleza, the other through Crateús. The local Indians, who in the interior tended to be hunters rather than agriculturalists, killed the cattle for food; and in the inevitable confrontation, they were eliminated (either killed, dispersed, or enslaved) as an independent force in the history of the area.

Cattle from the uplands acted as draft animals in the sugar fields and mills, and supplied the dried meat called "charque" or "carne do Ceará" to the markets on the coast. They were raised on large fazendas owned by absentee landlords and overseen by a vaqueiro, or perhaps by a trusted slave. The vaqueiro, if there was one, cared for the cattle, and was rewarded with one-fifth of the newborn calves each year. It was a position of prestige, and the average vaqueiro could eventually earn enough to purchase his own land and enter the ranks of the fazendeiros. Throughout the eighteenth century, the economic structure of Ceará did not change: beef and hides were the major exports, and agriculture, where it did occur, was largely for subsistence.

Cotton was well known to the Indians, and was among the first crop samples sent back to Portugal during the conquest. But until the last quarter of the eighteenth century, it had always been of marginal importance in the economy of Ceará. With the increasing demand for cotton that followed the dramatic expansion of the British textile industry at this time, however, cotton took on an importance in Ceará that it has never completely lost (Simonsen 1962: 370n). How cotton came to surpass cattle as the most important product is an interesting story; and in it we see two forces that have continuously shaped the agrarian life of Ceará. Cotton first gained importance in 1777, when the disruptions caused by the American Revolution forced its price up. Some years later, the severe drought of 1790–93 killed nearly all the cattle of the region. So we see from an early time that both the endemic droughts and fluctuations in world market prices were to be forces of major importance

TABLE 1
Summary of Cotton Exports Through
The Port of Fortaleza

Year	Kilos	Year	Kilos
1845	124,757	1910	3,043,250
1846[a]	46,378	1915[a]	4,929,230
1850	717,293	1922	16,005,368
1875	3,505,580	1930	16,107,100
1878[a]	628,948	1932[a]	4,089,091
1880	2,071,625	1935	25,000,000
1895	1,835,555	1940	28,000,000
1900[a]	2,008,330	1942[a]	19,047,434
1905	4,243,350	1944	30,000,000

Source: Girão 1947: 218–20.   [a] Drought years.

in the economic life of Ceará. Table 1 gives some idea of the impact that periodic droughts have had on cotton production in Ceará.

By the early 1800's, the salient characteristics of the semi-arid Northeast of today had already emerged. Many changes have occurred, however, without altering the basic structure. New strains of cotton have been introduced (Girão 1947); the use of slave labor has been abolished; and the average size of landholdings has decreased (Andrade 1960: 19). Perhaps the most striking development has been the steady replacement of cattle raising by agriculture, a change that is still continuing. This has produced an expanding population of subsistence-oriented peasantry with no interest in large-scale cattle raising.

Ceará today has a lesser concentration of landholdings and a greater diversity of agriculture than any other state in the Northeast. During the 1960's, Ceará was the number one producer of maize and beans in the region, and the number two producer of rice (Andrade 1960: 119–20). Agriculture, especially of the market-oriented kind, has expanded most thoroughly into the humid valleys and high serras, where the percentage of landholdings of less than ten hectares (minifundia) is around 30 per cent. In the less humid sertão, large landholdings concentrating on cattle, and subsistence agriculture, are more common.

The population of Ceará is still growing. Although the number of inhabitants per square kilometer is 22.6, about 34 per cent of this population is in the cities, particularly in Fortaleza; the population density of the rural sections (*municipios*) is frequently under 10 per square kilometer. The two most noticeable demographic trends are movement to the cities and movement out of the state. The first trend is a recent phenomenon: 77 per cent of the population was rural in 1940; in 1950, the figure dropped to 75 per cent; and in 1960, to 66 per cent. The second trend is older. Migration from the Northeast to other parts of Brazil has long been an important source of manpower for the successive "boom" economies based on mining, rubber, cacao, coffee, etc. It is estimated that the Northeast lost about 640,000 inhabitants through migration in the 1930's, and another 940,-000 in the 1940's (Baer 1965: 173).

Other changes have taken place within the last generation that are revolutionary from the traditional standpoint. Medicines are increasingly available to the backlanders; motor transportation is of growing importance; and radios, virtually unknown 20 years ago, are now commonplace everywhere. But Ceará is painfully backward when compared with any developed region. There are only 1,467 kilometers of functioning railroads and 350 kilometers of paved highway in an area of 149,895 square kilometers; and in 1960, the available electric power was only 10 watts per capita. The simplicity of the technology goes beyond these rather gross indicators, however. As late as 1960, there were only 316 tractors and 1,305 plows to a rural labor force of 816,720. Perhaps most telling of all is that Ceará's food production, except in fruits, does not equal its needs, forcing the importation of staple crops (Girão 1947: 458; IPE 1964: 194).

Two qualities of Ceará's economic life distinguish labor conditions there from those in many other plantation areas. The first is that Ceará is by no means overpopulated; labor shortages often occur, even in favored (i.e. humid) areas (Nicholls and Paiva 1966: 35), giving the laborers a degree of independence unknown to workers in a country of great land scarcity.* The second is the possibility of land being owned, albeit in small parcels of low quality, by many of the more successful workers. In fact, as the settlement of the state progresses, the number of small landholdings is increasing. Between 1950 and 1960 the total area under cultivation doubled, whereas the average size of agricultural landholdings decreased from 117.7 to 92.8 hectares; and there are some 20,000 farms of less than 2.6 hectares (IPE 1964: 178–79).

It is true that the smaller farms in Ceará do not give much economic security to their owners. "The small landowner is practically no better off than a sharecropper. He has, though, a moral advantage: he owns his house and land, and for this reason enjoys a sense of liberty. Yet from the financial point of view he is like a sharecropper, for a small holding in the drought region does not have the economic power it has in the South [i.e. São Paulo state].... The large landlord, under the system of exploitation in use in the region, rarely sustains a loss." (IPE 1964: 189.) Nonetheless, we can see that the average worker in Ceará has a strong bargaining position in his relations with the landlord: he can usually leave and find another plantation on which to live and work; or, if he has some capital, he can buy his own plot.

The economic life of Ceará centers around agriculture and commerce, and industry accounts for only 10 per cent of the

---

* Migration, rather than a low birthrate, is the cause of this mild labor shortage. Baer (1965: 173) demonstrates that the constant outflow of labor from the Northeast "has prevented the northeastern population from growing at an explosive rate in relation to its meager resources." And Robock (1963: 11) remarks that agricultural resettlement projects deep in the interior of Brazil also help draw off surplus farm population from the Northeast. The labor shortage in the vicinity of Boa Ventura may also have some connection with the simplicity of the technology (see Chapter Four). The labor shortage in Ceará must not be exaggerated, however, for B. J. Siegel reports a definite land shortage on fazendas in the Municipio of Quixadá (Siegel, personal communication).

income in the state. Cotton is the principal product, accounting for 12.8 per cent of the gross product. The agricultural system, at the most general level, is a complementary exploitation of cattle and cotton. The cattle forage among the cotton plants and the stubble of the subsistence crops planted with the cotton during the dry season, when wild grasses are no longer available; in turn, they fertilize the field, sustaining the cycle for many years.

Although the Ceará backlander often has Indian ancestry, he has no great weight of Indian traditions behind him, having inherited his language, his religion, and, indeed, most of his culture from the Portuguese. He is essentially pragmatic, aware that his world is rapidly changing and seemingly able to accept this change. Of course, he has traditional answers to many questions, but his ideas are in a state of flux. Unlike the narrowly bounded peasant communities of Highland South America, with their distinctive dress styles and their suspicion of strangers, the open community of the Brazilian backlander is remarkably receptive to outsiders. The moradores wear modern dress and decorate their homes with pictures from glossy national magazines. They circulate from fazenda to fazenda, and sometimes to urban centers, lacking roots in any closely confined area. Workers enjoy talking about themselves, much to the advantage of the ethnographic fieldworker.

In the class system that exists in the region, the worker from the interior (*matuto*) is easily recognized by the sophisticated: his clothes are of cheap cloth, and are sewn by local seamstresses; and his speech is poorly enunciated and ungrammatical. But although he may appear rough, his life is peaceful. Courtesy is elaborate and carefully observed, liquor is avoided by most workers, and open fighting is uncommon. Both men and women spend their days at work, and their evenings, which are short, either sitting quietly at home or visiting neighbors.

Three
Settlement and Migration

Much of the economic life of a resident of Boa Ventura is structured by the rules, or ideals, of settlement and migration that he follows. The pattern of residence within the fazenda influences where (and consequently what) he plants, and how easily he has access to fields, water, and shops (*bodegas*); and it determines the people with whom he forms his horizontal exchange ties. Likewise, the pattern of migration fundamentally influences the range of choices available to a morador. But our analysis must not stop at the rules by which workers make decisions affecting their moves and settlements, for these rules are often adaptive responses to certain basic and unavoidable environmental realities.

SETTLEMENT PATTERNS

During the period of this study, although there was much coming and going of families, the population of the Fazenda Boa Ventura stayed at about 305. This population is divided evenly between the sexes, indicating the importance of families, as opposed to single males (or females), in the work force of the plantation. Over half the residents are under 14 years of age (see Figure 2). This predominance of young people is maintained by a very high birthrate, and by a high survival rate after the first year. For the married women in my sample, I recorded 399 live births, or an average of 0.56 births for every year of

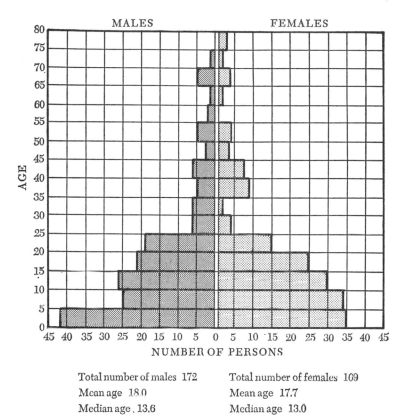

Total number of males 172   Total number of females 109
Mean age 18.0   Mean age 17.7
Median age 13.6   Median age 13.0

Figure 2. Distribution of population by age and sex (N = 341).

married life. Most deaths among children occur during the first year, giving Boa Ventura an infant mortality rate of 223 per 1,000 live births.* Of those who survive the first year, however, only 7 per cent will die before reaching maturity. It is not surprising, then, that families are large.

The average size of a nuclear family is 6.4 persons. This figure, it is true, does not include others who may be living under the

* Cf. the figure for Fortaleza, 166.7 per 1,000 in 1965; and for São Paulo, 69.9 per 1,000 in 1963.

same roof as the head of the household and his family; but the average household size, which is 6.8 persons, gives some indication of the relatively small place that non-nuclear kin have in these households.

Now let us look more carefully at the types of household that do occur in Boa Ventura, and at the patterns of settlement that characterized it at the time of this study. I will first present the facts of residence, and then discuss these facts in relation to the workers' ideals of residence and to the physical environment.

Of 51 households for which I have data,* 36 (or 71 per cent) are either nuclear or subnuclear. In this case, subnuclear refers to households of single persons, and to broken homes (e.g. a widow and children). The exact breakdown is: one widow with children, three one-person households, five married couples without children (either just married or so old that all their children have left), and 27 married couples with children. There remain 15 households of a more complex nature. Such households are usually very unstable, always gravitating toward the nuclear type in the course of time. I have distinguished three broad types.† In the first type, one of the marriage partners has some children by another marriage (or born out of wedlock) living in the household. Since the death of a spouse is not unusual, and remarriage shortly after the rule, this type is fairly common (six cases). The second type of complex household includes the sibling of one of the marriage partners, or some of the sibling's children, but not both; that is, when a person has the children of a sibling in his house, that sibling is always absent. (Because the fazenda provides housing, there is no reason for a sibling with children on the same fazenda to share a household.) The frequency of siblings or their children living with nuclear families (seven cases) can be explained largely in terms

---

* This exceeds the number of households present at any moment (about 45), because some houses were tabulated before the members moved away and again when a new family moved in.

† These types are not particularly significant to the residents of Boa Ventura, who consider any non-nuclear household an exception.

of labor needs, since all cases but one involve able-bodied young men or women. In the third type of household a non-relative is living temporarily in the household for some reason, usually as a hired laborer. This circumstance usually occurs in a household that is already complex in one of the first two ways, but two households are complex in only this way.

The physical placement of households extends to the boundaries of the fazenda; a worker does not mind living at a distance from other workers—in fact, a certain minimum distance, something over 100 meters, is absolutely essential for good neighborly relations, as we shall see momentarily. Aside from the moradores who are settled at the gates opening out of the fazenda onto the main roads, however, most households are clustered in groups very much like neighborhoods. There are six such neighborhoods on Boa Ventura, each with its own local name.* Within a neighborhood, houses tend to be built on either side of some well-traveled path, and none are built more than a few dozen yards away from such a path (see Figure 3).

Neighborhoods differ greatly in their access to farmland, water, shops, pasture for donkeys, and other valued goods and services. A neighborhood also has a more subtle quality arising from the presence or absence of kinsmen, and from the predominating lines of local exchange. As we shall see later, moradores have some choice in where they live, and all of them have definite preferences. Particularly noteworthy is the tendency of kinsmen to cluster in one neighborhood. Not all families have close kin living on the same plantation; but when they do, their kin live either next door or at least within the same neighborhood, with very few exceptions.

Two moradores on Boa Ventura have horses, and a few others have donkeys; but for most, transportation is by foot. Within neighborhoods, distances are not at all great, and the largest contains no houses more than 15 minutes' walk from one an-

---

* There are eight named sections of the fazenda, but two of them have only one resident family.

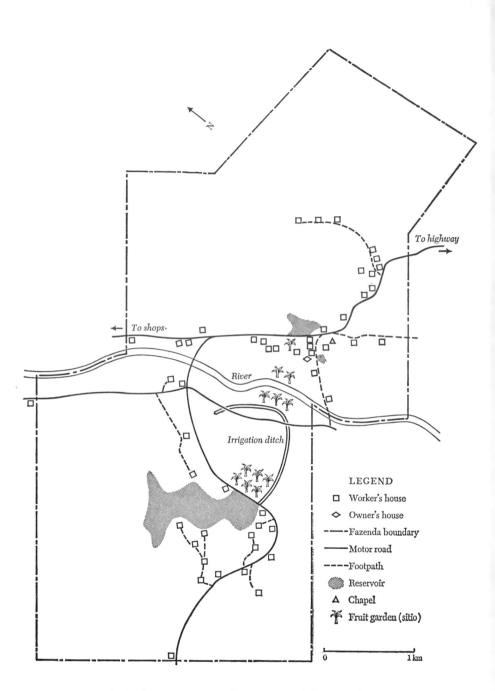

Figure 3. Settlement pattern and major physical features of Boa Ventura.

other. But the neighborhoods are separated by relatively large uninhabited stretches, and the distance between them is always more than 15 minutes' walk. All households are within easy walking distance of some small shop where emergency purchases such as matches, kerosene, or salt can be made; but access to the trade centers where staple purchases are normally made varies greatly. One isolated neighborhood is two hours' travel, by horse, from the cluster of shops where the residents do their major purchasing. For men who work practically from dawn to dusk six days a week, distances of over an hour's walk are prohibitive; hence the location of a household greatly influences visiting and shopping behavior.

As I have already mentioned, the nuclear family is the typical residence unit, and more complex units tend to disintegrate through time in the direction of the nuclear family. This is chiefly a result of the ideals of settlement that the moradores hold.

The primary ideal of a married couple is to be the *donos de casa* ("masters of the house") of their own dwelling. The term *donu de casa* implies control over the resources of the household, and applies as much to the wife as to the husband, although the two have distinct spheres of economic concern. It is inconceivable to the moradores that any household could have more than one dominant couple, and any married couple expects to exercise full control over its own set of resources in labor and property. Thus any household containing more than one married couple presents a fundamental contradiction.

In practice, the only households with more than one married couple are those in which newly married children have not yet been able to set up their own household. This situation is always temporary: a new house will be built, a suitable dwelling will become available, or the young couple will simply move to another fazenda. In any case, the complex household soon becomes a nuclear family. The households that include an unmarried relative are far more stable, for then there is no challenge to the couple's rights over household resources, including the labor of the additional member; in fact the newcomer's labor is often a major resource, as in the case of an unmarried man living in the household of an older brother who has a large family but no grown sons.

The second important ideal of residence is to live near people with whom one can have, at the very least, peaceful relations, and at best relations of mutual aid. This ideal is greatly served when the third and fourth ideals are fulfilled.

Third, it is generally agreed that it is a good thing to live near one's close relatives. Living nearby and living in the same household are two very different things, and the claims that neighboring kinsmen can make on each other are far less than those that a dono de casa can make on members of his own household. The demands made on neighboring kin are the same as those made on any neighbors; however, kinsmen are much more likely to fulfill them than non-kin. Neighborliness (making tea for the ill, making small purchases in the shops, extending small loans, and so on) is a valued relationship; and it is far easier to form these relationships with kinsmen.

The fourth ideal of residence, never ignored as far as I know, is that of a minimum distance between households. To the moradores, the physical proximity of neighbors is not a source of security but rather a source of tension. Most of this tension derives from household livestock. If houses are too close together, pigs and chickens may dig up a neighbor's yard, or take feed meant for his own animals. Children, who are taught respect for the property of others only in a leisurely fashion, can also provide considerable irritation. The moradores are very intense in their feelings about this, and I noticed conflicts even between neighbors living a hundred meters apart. In one case, two families shared the same fenced yard for a few months while another residence was being installed; both parties found the arrangement practically intolerable and complained bitterly. Living next to relatives helps to reduce this source of conflict, since kinsmen are inclined to see themselves as part of a larger, if loosely formed, unit, so that the losses caused by marauding pigs or chickens are not felt so strongly.

Finally, a high value is placed on companionship. Moradores like isolation; but they also like to have access to other people, and particularly to be near the favorite gathering places of the fazenda (cf. Harris 1956: 32).

Certain physical features of a homesite are considered desirable by the moradores and have an effect on where they finally settle (insofar as they are given a choice). First, and apparently most important, the house should be located near a year-round source of water. Second, a house is preferably built by the morador himself, or at least on a site that he selects. In this way, he is assured that the house will be sturdy and will have the space he requires, including storage space for his harvest (at least a quarter of the moradores have achieved this ideal). Finally, a house should be near good land. This ideal is flexible, since most moradores expect to walk a considerable distance to get to their fields; nonetheless, workers from one neighborhood very seldom farm land in an area predominantly worked by residents of another neighborhood.

In practice, workers are usually able to settle in accordance with their ideals. Moreover, these ideals are quite sensible formulations of the workers' adaptive responses to economic conditions. The spacing of the houses shown in Figure 3, for example, may be largely accounted for on this basis. The two largest neighborhood clusters are near the two largest reservoirs, and have easy access to water, shops, and companionship. Most houses are separated from others by at least a hundred meters, but are still within a few hundred meters of at least one other house. This represents a compromise between the need for space to raise livestock and the need for easy access to an exchange partner. The ten houses in the northeastern neighborhood are inconveniently placed relative to shops and water; but they are compensated by easy access to the largest and most fertile hillside agricultural plots.

MIGRATION, DROUGHT, AND THE LABOR SUPPLY

One of the most striking aspects of settlement on the plantations of Ceará is the instability of the farm population. This has been

TABLE 2
Migration on Boa Ventura

Number of Moves	All Moves Between Fazendas		Between Fazendas (Different Municipios)		Within Boa Ventura	
	Households Moved[a]	Total Moves	Households Moved	Total Moves	Households Moved	Total Moves
0	11	0	23	0	24	0
1	11	11	12	12	12	12
2	13	26	2	4	2	4
3	3	9	1	3	2	6
4	1	4	1	4	0	0
5	0	0	0	0	0	0
6	0	0	0	0	0	0
7	0	0	1	7	0	0
8	1	8	0	0	0	0
Totals	40	58	40	30	40	22

[a] To compile these figures, I asked the heads of households to discuss the times they had moved since they began housekeeping. The figures are for married couples only, and do not include moves by adults before marriage. Thus zero moves in the Moves Between Fazendas category means that an individual worker came to Boa Ventura single, was married there, and has not yet moved away.

particularly evident in migration out of the region, caused in part by the periodic catastrophe of the drought. In the nineteenth century, it was proverbial that the northeasterner had an "instinct of migration," which impelled him to move after a certain length of residence in one spot. Indeed, there are constant changes in residence by the workers, and drought is by no means the only reason for this circulation.

Movement from house to house within Fazenda Boa Ventura is common, though not as common as migration in and out of the fazenda (see Table 2). This internal migration is partly a response to movement away from Boa Ventura, for as houses become vacant people in inferior houses move to the choicer locations. During the nine-month period between October 1966 and June 1967, in an average population of 305 persons, 35 persons (11 per cent) in five households moved away, and 34 persons in six households moved in. Another six households moved from one house to another within the plantation. In October 1966, there were 46 households; in July 1967, there were 47. Three new houses were built, and three old ones were destroyed.

While waiting for houses, the newly arrived moradores made various temporary arrangements.

The fazenda manager, to keep the system working smoothly, often moves families, with or without their agreement, to a position more desirable from the plantation point of view. He may move a small family out of a large house and move a large family in, or he may move a family out to the boundary of the fazenda to watch the gates. For whatever reason, almost all families that have spent more than a few years on Boa Ventura have moved or been moved once; at the time of this study, the longest residence in a single location was 13 years.

In the following discussion of migration within Ceará, it should be emphasized that the sample with which we are dealing is of a particular sort; that is, it must be contrasted with a sample that might be drawn from workers who have migrated from Ceará to the South (e.g. Rio de Janeiro) and stayed there. To my knowledge, only three or four men on the plantation had ever been outside the state of Ceará in their lives. It could be argued that the type of worker who leaves his homeland and stays away, perhaps adapting to very different social and economic circumstances, is likely to have a very different personality from the worker who has never even attempted such a move. Similarly, the kind of migration with which I am mainly concerned—namely, migration from fazenda to fazenda within the same social and economic setting (in fact, within the political borders of Ceará)—could easily differ from migration out of the state, which has been so dramatic in the past and has excited so much interest in the literature.

Most of my discussion here is based directly on my informants' explanations of why they made each move they did. There are, however, a few generalizations that can be made, of which the moradores themselves do not seem to be aware.

First, about half (28 out of 58) of the moves from one fazenda to another were made between adjacent fazendas, or at least between fazendas in the same municipio. Relatively distant places like Sobral, Redenção, or Fortaleza were mentioned in the interviews, but they were unusual.* In general, moradores tended to move to a place with which they were already familiar, perhaps through personal knowledge of the landlord or manager, or through the presence of kinsmen. Migration is a commonplace experience for the moradores, but it is not undertaken lightly. On the contrary, it usually involves careful planning so that both landlords involved will be prepared for the date of departure or arrival. And since the security of his family is at stake, a morador does not leave one plantation without having established his place on another.

Another general characteristic of migration is the tendency for couples to move to another plantation during the first year of their marriage. There is a norm of neolocal residence at marriage, and it is a strong norm. In many cases, a couple is married at a time when there are no vacancies on the fazenda where they are living. In this situation, they can usually find a place on a nearby plantation; or, if the husband was a migrant worker, they can rejoin his relatives on some distant plantation.

A final tendency apparent from my data is that a worker is not inclined to return to a given plantation once he has left it. There are a few cases to the contrary, but out of over a hundred moves recorded, less than ten were back to a previous plantation. Considering the backlander's preference for the known versus the unknown, this is rather mysterious. The workers themselves can offer no explanation. However, there are a few obvious reasons for this behavior. First, leaving one plantation for another is an open admission by the worker that his access to desirable land, water sources, or convenient shops was not enough to satisfy him. This being the case, why should he go back? Second, there are relatively few open positions on a given plantation at any one time; the decision of precisely which plantation to move to is partially determined by the chance openings existing at that moment, and one is not always free to return to

---

* Workers travel relatively far within the state to visit kin; however, these visits will be made only once a year, or less often, and for short periods of time.

TABLE 3

Inter-Fazenda Migration

Reason for Move	Percentage of Total Reasons Given (N=67)
1. Left in search of work during drought	15%
2. Came by invitation of landlord	14
3. Left because of shortage of good land	12
4. Left to live on small landholding owned by relative	12
5. Left after a quarrel with the landlord or manager	10
6. Came because a close relative was already a resident and recommended the place	8
7. Left to follow a trade in more promising circumstances	6
8. Left because landlord's conditions were too heavy an imposition	4
9. Came because he heard conditions were better, by rumor	4
10. Various other reasons	15
Total	100%

a former home. Finally, workers often leave because they have been insulted, and any move away from a plantation implies some criticism of the landlord or manager. For these reasons, a worker may leave behind a certain amount of bad feeling that would make future cooperation difficult.

Now let us turn to the moradores' own reasons for migration. I collected statements giving the reasons for 67 of the moves from one plantation to another, as given in Table 3. This list, however, does not indicate the full complexity of the normal decision to migrate. It is probably rare for any single reason except the drought to prompt the decision to move by itself. For example, Reason 9, that the morador left because rumor had it there were better conditions elsewhere, implies that conditions at his own plantation had become unattractive, perhaps for one of the other reasons (e.g. because of a shortage of good land).

Quite properly, the most common reason given for moving was the worker's inability to subsist in a certain location during a drought year. Actually, the figure of 15 per cent given for this response may underestimate the importance of the drought as a stimulus to migration. Of the 103 total moves I recorded, 24

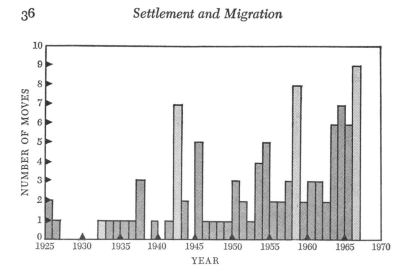

Figure 4. Drought and migration. Drought years were 1932, 1942, 1958, and 1966.

per cent took place during the drought years of 1932, 1942, and 1958, and the near-drought year of 1966 (see Figure 4). In many cases, the drought, which affects those with the most marginal relation to the land the most severely, may have finally decided people who were already in a mood to move away but had so far hesitated.

In Ceará, a true drought is not just an additional pressure that acts as a sort of malthusian check on a delicately balanced population-environment relationship. It is a catastrophe of enormous magnitude, quite different from the already extreme variations in rainfall. The population can suffer through a relatively dry year, and will grow fat in a good rainy year; but it cannot by any means be supported solely by agriculture during a drought. Nonetheless, the importance of the drought as a direct influence in the decision to migrate should not be overestimated. There are, and have been for some generations, methods by which at least a part of the population can escape the worst effects of the drought and remain in the same locale until the rains begin.

For one thing, there are the reservoirs, which have been built over the last century or more to provide water for both cattle and people during the droughts. The level of water in the reservoirs shrinks as the drought progresses, exposing successively richer strips of wet land; these can be planted in food crops by workers who have access to them, usually through close ties with a *patrão* (patron). Also, the Brazilian government has recently been investing large amounts in public works, such as road building and reservoir maintenance during the droughts. Thus workers who cannot farm may still earn a livelihood without having to travel very far from their homes. If a worker has a "strong patrão"—that is, a wealthy landlord—he may be able to do work such as repairing reservoirs, clearing stumps from fields, or building irrigation canals for wages (often paid in food and cloth), and can thereby outlast the drought without moving from home at all. Finally, if a worker has taken the care, as many do, to build up a surplus of grain the year before the drought, he may be able to remain where he is, living off his surplus and working for wages whenever possible.

Thus the drought implies a reduction, but not a reduction to nothing, of the backlander's economic resources. Marginal resources, or resources normally employed in other sectors of the economy (e.g. the savings of a landlord, which are often invested on the plantation during droughts in order to retain favored laborers) are available to help protect at least a part of the peasantry from the full effects of the drought.

Putting aside Reason 2 for a moment, the third reason given for migrating was that there was a shortage of good land where the worker had been living, and that he knew of a place where better or more plentiful land was available. This reason is a very common one because the state of Ceará is still being settled by an expanding rural population. As population pressure builds in one area, some labor is inevitably attracted to less populated areas.

There is nothing very surprising about the appearance of drought and land shortage as two important variables in migration in Ceará. In most standard treatments of population dynamics we find migration considered as a redistribution of the human factors of production to meet differential economic characteristics of the environment (Beshers 1967: 142–45; Thomlinson 1965). These reasons, however, accounted for only 27 per cent of the explanations offered by moradores for their moves. Let us now consider some of the other reasons given for making a move, and it will become clear that they are to be explained quite differently from the reasons of drought and land shortage.

The second most common reason given for moving was that the new landlord had invited the morador to migrate. From interviews, I learned that several moradores on Boa Ventura, although they had no definite plans to migrate, had standing offers from other landlords in the vicinity, to the effect that there would be a place for the worker on a neighboring plantation should he ever want to move. This and other evidence points to a certain competition, mild though it may be, among landlords for labor resources. Labor has historically been scarce in the Northeast, particularly in the more arid regions. Since the abolition of slavery, the laborer has always had this in his favor in economic transactions with the landlord. The present-day morador is aware of this, and expresses it in his feeling that he serves the interest of the landlord by providing his "arms" just as the landlord provides land: "The landlord does no work; without our arms he would earn nothing."

Any given landlord has complete control of the land from which a morador must earn his living; but other landlords can usually make a place for the worker on their lands, which greatly diminishes the repressive power that would ordinarily accrue to the landlord as a controller of vital resources. In the view of the morador, his relations with the fazenda owner are a give-and-take, and the owner is expected to act honestly and respectfully toward him.

The relationship between the worker's sense of independence, the labor shortage, and migration is strengthened by the following fact: there is no evidence that a seasonally migrant labor force exists. During heavy work periods, such as the start of the rainy season, wages climb to double their previous level. This indicates that the labor supply is relatively inelastic, forcing wages up with an increase in demand. If there were any significant "hidden unemployment," such as is claimed for many overpopulated regions, one would expect the availability of labor to increase along with the increase in demand (see Johnson and Siegel 1968).

All this gives the moradores of Ceará a remarkable independence and freedom of movement. In this regard, it is useful to note that reports of bad faith on the part of landlords are very infrequent. A landlord could at any time evict a morador and refuse to reimburse his labor, and the morador, poor and illiterate, would have no recourse; yet only one case of this was reported to me. Other moradores reported hearing of such injustices, but had never experienced them themselves. As far as I can tell, a landlord is very careful not to get a reputation among the moradores that would discourage potential laborers or send his own moradores in search of a more reputable patrão.

Other reasons commonly given for migrating make sense in the above terms. That a morador may be invited to live on a relative's landholding (Reason 4) arises from the circumstance that many poor peasants do manage to purchase a small plot of their own farmland. The advantage to the worker who goes to live with, say, an uncle on such a plot is that he seldom has to give the shares of his produce he would be required to give on a large plantation. On the other hand, the plots owned by moradores are usually too small and too poor in quality for him to do better for himself than he can on a plantation, landlord's shares or not. The remaining reasons for migration (6–9) all reflect a worker's dislike for some aspect of one plantation or his attraction to some known situation elsewhere. Moving after a quarrel with a landlord or manager is common, and normally follows some insult to the pride of the worker; indeed, the pride and swagger of the backlands worker is at times more reminiscent of the Bedouin than of anything that one associates with the word "peasant."*

My remarks so far have implied that a morador has no ties or encumbrances that would interfere with his freedom to move when circumstances become distasteful to him. In general, this is true. A worker's household possessions are few; after long residence he may have acquired things like tables and benches, which he cannot move very far, but these are easily sold. His kin ties involve no long-term obligations, and his "contract" with the landlord, such as it is, never prescribes the length of his stay.

There is, however, one major restraint on a morador's freedom to leave the fazenda. During his residence on a plantation, a worker usually establishes fields of cotton, and sometimes fruit trees, which will give returns for years with a relatively small input of labor, once the initial chore of clearing the field and planting the crop has been done. As I will show in Chapter Four, such labor investments are quantified and given money value by the peasants, and they can be sold to other moradores or to the landlord. Ideally, then, a worker can leave the plantation without losing the long-term fruits of his labor simply by selling future rights to the produce of his fields. When a morador suddenly leaves for his own reasons (without being put off by the landlord), however, the landlord will practically never purchase those rights. Hence a worker who cannot find another morador to purchase his producing plots must leave without compensation, the rights falling by default to the landlord.

It often happens that no other moradores have the cash or the desire to purchase the crops of one who wants to leave, and this may be a very effective restraint on a worker's mobility. Any morador who has resided in one place for more than two years will have at least a few fruit and cotton plantings, and the longer his residence, the greater his investment in labor (up to a point). The restraining effects of this factor, then, will be felt by

* Reason 7 concerns skilled labor, which is rare. On Boa Ventura, this involved only one man, who moved frequently and always for the same reason—to improve his chances for practicing his trade.

nearly all workers except the most recent arrivals on a plantation. When this fact is coupled with the average morador's view that conditions on one plantation are for the most part just like conditions on the next, then what stability there is in the sertão population is largely accounted for.

What does migration within Ceará seem to accomplish? The data gathered in this study are not sufficient for a complete answer to this question, and larger demographic processes may be taking place than those measured here. (For example, the interviews give no hint that a general trend of migration into Fortaleza or the great metropolitan centers to the south exists; yet we know from other sources that this process is going on.) We do know that during a drought year a good proportion of the moradores may move from one fazenda to another; moreover, it is clear that they move because resources in one area have become insufficient, and that they believe adequate resources are available elsewhere. This has an "economic" implication in that a certain factor of production, human labor, is being redistributed in accordance with the availability of other factors of production.

In general, there is no question that a labor market is functioning here to redistribute labor from areas of low demand to those of higher demand. But another principle seems to be operating as well. When a morador leaves a fazenda after a quarrel with his landlord, or because he was invited elsewhere by another landlord and wants to try his luck, he leaves a gap that will be filled by another morador possessed of approximately the same rudimentary skills as the worker who left (except in the case of the specialists discussed in Chapter Five). The net "economic" effect, in terms of anything measurable, is effectively zero, for one quantity has been replaced by an identical quantity. Nonetheless, something important has happened: the morador has asserted his independence; he has drawn the line at which his dependence on a particular patrão ends. He could do this only with great difficulty, if at all, in a region where population pressure severely limited the availability of land.

Four

The Subsistence Economy: Agriculture

The day-to-day productive activities of Boa Ventura are undertaken by moradores as members of individual households. By and large, it is the male head of the household whose agricultural labor supports the other members. In this chapter, besides describing how labor is allocated, I will outline the worker's concepts regarding land, crops, and agricultural techniques, and will explore the planning and decisions he bases on these concepts. To some extent, I will also attempt to relate those concepts to observable material conditions. In effect, this is a way of asking whether the workers are making good use of their technology and environment, and whether (and how easily) improvements might be made.

The reader will notice several examples indicating that most of the moradores plan their work carefully, calculate the respective values of alternative means and ends, attempt to minimize potential risk, and make decisions not only by conventional rules but also on the basis of their own observations and experiments. Thus a worker plans ahead so that the various factors of production (e.g. labor or work animals) will be available when needed; he invariably compares the value of a man-day of labor in his own fields with the value of a day spent working for wages; and he plants a wide variety of crops in a wide variety of micro-environments to minimize the effect of erratic weather. Moradores will generally follow tested planting methods, but each will plant according to his personal assessment of the different methods; and many will experiment with new or unusual crops.

Nonetheless, much, perhaps most, of the subsistence activity on Boa Ventura is carried out according to shared understandings from which individuals rarely deviate. That is, though two men may reach different decisions, the framework in which the decisions are made is usually the same; for example, two men may agree that a fertilized plot will yield more potatoes, yet for other reasons only one of them actually uses the fertilizer. The concepts of land tenure, outlined at some length below, are widely, if not universally, shared; thus land rights are used and transferred with a minimum of conflict or misunderstanding.

LAND

The moradores of Boa Ventura have an elaborate knowledge of the lands available in their environment, and make many distinctions that influence their productive activities. These distinctions are of two kinds: those between qualities of soil, and those between uses to which the soil is put.

As mentioned earlier, a basic fact in this environment is the variability of rainfall; hence the moradores categorize land chiefly according to its water-retaining capacity. Lands that retain water ("cold" lands) are best in a dry year, whereas lands that drain well ("hot") are equally valuable in a wet year. In general, "hot," dry land is found on the hillsides, and "cold," wet land in the low areas; but even in the hilly fields, clay soils, which do not drain as well as the sandy soils also found there, are considered "cold." Cold and hot soils, of course, must be planted differently, and they may occur right next to each other in the same clearing; so a worker must be well aware of soil distinctions if he is to work his land efficiently.

In general, the better lands in the sertão are low. The best land on Boa Ventura is found along the margins of the river (*coroa*), where periodic flooding replenishes the topsoil with silt. A similar effect, though less potent, occurs on the margins of the reservoirs (*vazante*), which are exposed by evaporation and seepage during the summer months, and in various limited areas where standing water collects in winter. It is easy to see that these qualities, basic as they may appear, are not constant but will vary with the amount of rainfall. For example, land that is only good for rice in a rainy year may support several other crops in a dry year. If a farmer could predict, even roughly, the amount and distribution of rainfall to expect, he could adjust his planting to take maximum advantage of this variability. As we shall see below, he does plan; but he does so to minimize the potential hazards of the climate, which is a very different matter.

There are also marginal or useless lands in this environment. But here, too, there is considerable variation, and the application of special techniques or the presence of unusual environmental conditions may open up or close down some areas. For example, seepage from a dam normally allows so much water into the area immediately below it that this area is too wet to work for most of the year; yet one morador, the only one with a plow, had opened up such an area by digging a drainage ditch. Another morador, gambling that those who were predicting an excellent rainfall were right, planted in a normally worthless piece of hillside (*campestre*), and was rewarded with a good season and a good crop. On the other hand, the desirable coroa land on the river margins may be suddenly inundated by unusually persistent rains or a broken dam, destroying the young crops (often, however, a new crop can be planted with a moderate chance of yielding). Similarly, lowlands that have consistently yielded well for many years may, if drainage is too poor, gradually become saline and infertile. And so on.

Workers on Boa Ventura also describe lands according to the use to which they are put. They distinguish woods and forest (*mata*) from cleared land, and completely regenerated forest from second growth. (This distinction is made at clearing time: the light clearing of second-growth *broca fina* is distinct from the heavy clearing of a regenerated forest, or *broca grossa*.) Any clearing that has just been cut and burned is known as a new clearing (*roçado novo*), and is distinguished from fields that have been under cultivation for a year or more (*capoeira*). This distinction is most important in the hill plots, where fertility declines sharply after the first year. A capoeira normally contains only tree cotton, although in the second, and even the third, year it may contain some corn and beans; but it never contains a full, first-year complement of crops. A plot that has all the stumps pulled up when it is cleared, so that a plow may be used (rare on Boa Ventura), is termed a *campo*. All the clearings mentioned are usually planted with mixed croppings; but a farmer will often set aside a plot of land (*lastro*) exclusively planted in one crop, particularly beans. Finally, any land that is at the moment being used as pasture for cattle may be referred to as *solta*, whether it is fallow, woods, or simply a harvested clearing where cattle have been let in to forage among the stubble.

Land tenure. There is no great land shortage on Boa Ventura, but there is an acute shortage of certain highly desirable land types. The best irrigated land is used by the owner for his fruit gardens (*sitios*). Land along the river is not available to most moradores. And in the most exploited sections of the fazenda there is a dearth of regenerated forest. With these limitations on access to land in mind, let us consider the kinds of land tenure encountered in this economic system, and the patterns of land allocation.

Boa Ventura is part of a larger economic system that includes small landholdings (*minifúndia*) as well as large, and five of the moradores living there own plots of land elsewhere. Three of these are homesites, but two of them are productive agricultural land. Four other moradores report that they are direct heirs to parcels of land (i.e. their parents or their wives' parents own land). Whether they own the land outright or can claim rights through filiation, the reasons workers give for not living on their own land are very similar. All of them have tried to live on the land; but when a bad year came they could not make ends meet and had a find a *patrão*, or too many relatives were trying to live off the same small, usually poor, piece of land.

In general, however, a morador hesitates to dissociate himself from even the poorest piece of land, since one of his major goals is to own his own land, if only as a symbol of independence. This drive is also apparent among the owners of homesites, who look forward to a more independent existence once they can gather the money to erect a house on their land (which is generally in some small village nearby). Many people do make this escape from the dead end of plantation life. Once free of the burden of rent, or the conditions imposed by a landlord for their use of a house, they may hire out as wage laborers to nearby fazendas, practice some craft, or seek the support of their nearby sons and daughters.

Some moradores living on Boa Ventura have at one time or another left a fazenda to go live in a village, paying rent for a house and selling their labor on the local market. But renting is a very uncommon transaction in this region of Ceará, and is not found at all on the plantations. Those who go to a village are usually discouraged by the uncertain labor market, the difficulty of paying cash rent, and the impersonal relations of village life, which make cheating more likely; they quickly return to the relative security of a plantation. Moreover, renting in Ceará is not entirely free of customary restrictions, and these may discourage prospective lessors as well as tenants. For example, one man who owns two houses near Boa Ventura refuses to rent them while waiting for a buyer at his price; he fears that a renter, once established and paying rent, will feel a continuing right to stay there.

The majority of the moradores on Boa Ventura own no land; they never have and never will. For them, tenure is a matter of rights to use the land of someone else. Usually, it is the fazenda manager who allots land for use when asked by a worker; but in special cases, such as bottomlands for plowing, fruit gardens, or fazenda crops of feed grasses and sugar cane, the fazendeiro himself makes the decisions. Often, in fact, the owner will offer a desirable piece of land without being asked to a worker who seems especially energetic or skilled; the aim is to get the best land into the most productive hands.

Every worker is entitled to as much hillside as he and the other able-bodied workers in his family can care for. Each year, he must clear a new plot, in accordance with the farming methods practiced. He acquires this plot simply by asking, subject to two limiting factors. First, because large, farmed-over tracts (*mangas*) are used for cattle grazing in the fallow years, and because the only fences are those between mangas, it is more efficient if all new agricultural plots are cleared in one manga, so that cattle may graze freely in the others without damaging crops. The second limiting factor has only appeared in recent years, as forest clearing outruns regeneration. So far, this has only happened in the most populous section of the plantation: here, the workers have not been able to clear new plots for two years because the landlord, fearful of destroying the fertility of the land by short fallowing, has ordered them to continue farming in existing plots.

A worker will usually farm a single large roçado, although some workers farm two or three. When more than one plot is farmed, it is usually because the available land near a worker's house is not fertile or extensive enough to sustain himself and his family. Some fields are naturally bounded by roads, creeks, or wasteland; but most must be marked off before they are cleared. Using natural markers like rocks or trees, workers confer with the manager and lay out their plots. This is all decided long before the planting season, generally in August or September. The plans made at this time are often inaccurate; owing to unforeseen contingencies like illness, some men are never able to clear all the land they have been assigned, whereas others can ask for and farm additional roçados. In general, the fazenda allows workers to clear as much land as they want (except in the heavily populated area mentioned above, where good roçado land is scarce).

Access to the more desirable land in the lowlands and along the river is obtained directly from the landlord as often as from the manager. This land, which has a much steadier water supply and more fertile soil, is an essential part of the workers' security system, and those who plant there are better protected against a poor growing season than those who do not. Two factors seem to govern the allocation of choice land. First, the worker himself must demonstrate his interest by asking to farm the land (surprisingly, not all workers seem to be interested, perhaps because they do not feel themselves capable of caring for anything but a roçado plot). Second, the landlord and manager must feel that the worker in question is both loyal to the fazenda and competent to farm the land efficiently. Prime land is generally regarded as a reward for good work and loyal tenancy.

There is never any doubt that the landlord has the ultimate right to dispose of the land. Nonetheless, certain conventions recognize the morador's temporary possession of the land he is working on. In all cases, these temporary rights are originally established and maintained through labor. The fundamental rule, subject to the limitations noted above, is that any uncultivated land is available for use by any interested worker. The converse, that any land currently under cultivation pertains to the worker on it, is carefully observed. In fact, a few moradores who have lived on Boa Ventura for many years have established and maintained uninterrupted tenure on certain pieces of land, and they are allowed to distribute these lands for the use of other moradores (usually relatives). At the same time, they are the men responsible for turning over an appropriate share of the produce to the landlord.

Very commonly, a worker finds that he cannot continue to care for a piece of land and stops farming it. Someone else may then point out to the fazenda manager that this land is degenerating and ask permission to work it for himself. Disputes over the eventual harvest are frequent in this case. The initial tenant, whose labor established some of the enduring qualities of that field—e.g. clearing and the planting of perennials like cotton, bananas, or manioc—will feel he still deserves some compensation. And, indeed, the succeeding tenant will usually settle with him for a payment in cash or kind. Although the first worker had abandoned the field and would have lost all return if no one else had taken it over, both workers feel that if any product is forthcoming, the first worker has contributed to its value and should be repaid.

The rights established through labor investment often vary in degree. For example, when the manager or owner, for some reason, wants to evict a worker, the owner is obliged to pay for the worker's cotton trees and any other improvements that will continue to produce after the worker leaves. This obligation holds when the worker has done all the labor of clearing, planting, and weeding. But if the worker acquires a producing field that has been abandoned by another who left the plantation of his own accord and does no more than weed and harvest the cotton, he receives nothing when evicted. The complete labor of establishing and caring for a field is viewed as entitling the worker to all the future produce of the plants (including the perennial cotton and manioc, or to reimbursement if the produce is taken from him; merely tending an existing crop will entitle him to that year's harvest, but he has gained no continuing rights. Thus the compensation that must be paid for a producing field taken over by the fazenda is not determined simply by the size of the field, nor by the expected value of its yield in the future, but also by the actual labor that a worker has invested in it.

An apparently contradictory example occurs when two workers have neighboring fields. Squash plants, which are almost always present in a mixed crop, often send trailers from one field into another. (Fields are usually rectangular, with their boundaries marked by a stick placed in the ground between the two fields.) As the moradores see it, the labor of clearing ground and planting the seed from which that plant grew has established a right to the squash. Nevertheless, if A's squash grows over B's field, B keeps it. This is not, as one might first imagine, because B has somehow gotten absolute rights to everything within a given territory defined by geographical limits. Rather, an extraneous cultural consideration has intruded: to take any produce whatever out of the field farmed by another is to risk being called a thief. In the backlands, acquiring a reputation for thievery, or even being suspected of it, is one of the

most destructive things that can happen to a man; it is a noto-
riety he can never live down. In spite of this, the right established by labor has not been forgotten by the moradores. As
one informant put it, *A* voluntarily cedes the squash to *B* in
order to avoid the dangerous suspicions that would arise if he
ever set foot in *B*'s field.

In sum, although a morador does not own land, he establishes
definite rights, which amount to a kind of tenure, by applying his
labor to the land.* Furthermore, these rights can be bought,
sold, traded, and given away, much like any other commodity
the moradores deal with. Such transactions are extremely common. A man leaving the plantation will try to sell his cotton
fields rather than abandoning them; likewise, a new arrival will
try to buy the rights to some cotton plants rather than wait for
several years to build up his own. There are also cases of men
trading fields to their mutual advantage. In these transactions,
as in most exchanges on Boa Ventura, the parties assess the
value of the field in terms of cash, even when actual payment
is in goods or land rights. A field's worth depends on the labor
put into it, figured by man-days, from which a certain amount
is subtracted to account for the age (i.e. lowered productivity)
of the field.

The transfer of land rights between moradores usually takes
place without involving the landlord or his manager;† the man-

---

* Salisbury (1962: 68) describes a slightly more complex system found
among the Siane of New Guinea. There, one can isolate three kinds of
rights to land: (1) the right of a lineage head to control the transfer of land
to those outside the lineage; (2) individual "ownership" of the land with
rights to sell or trade, provided the lineage head approves; (3) rights to
"the increment to the untilled soil caused by labor." The last is identical to
the tenure system on Boa Ventura. In the terms of the Western economist,
this practice distinguishes between land assets and assets that are improvements on the land. At the same time, the situation on Boa Ventura is a system of usufruct rights to land, commonly seen by anthropologists as a kind
of land tenure; in this regard, questions of how rights are acquired and disposed of are particularly interesting. For further remarks on traditional
swidden agriculture and land tenure, see Conklin 1957.

† Conflicts with the manager over land rights are virtually nonexistent.
Conflicts between moradores are common, however, and the manager may
be asked to intervene by one or both of the parties involved.

ager may be notified after a transaction is completed, but no great importance is attached to doing this. The transfer of any significant rights generally involves payment, but minor rights may be presented as gifts to workers who have a continuing exchange relation with the giver. Some of these gifts are no more than rights to small, undeveloped portions of a landholding; but I encountered two cases involving more substantial gifts. The first was a well-developed section of a fruit garden, given by a worker to his son-in-law. The recipient gained the right to tend the land for one year and dispose of that year's crop; but his father-in-law expected to reclaim future rights for himself.

The second case is an unusually clear example of the relation between labor input and tenure. A young bachelor had lived with his father for several years but had worked separate fields of his own; as was customary, the total product of the son's labor belonged to his father. When the young man was married, however, he was given rights to the cotton fields that he had established by his own labor and to no others. The "gift" amounted to nothing more or less than the complete transfer of the son's labor investment in cotton fields from the father to the son at marriage.

Use of land on Boa Ventura is not restricted to the resident moradores. In the year of this study, eight men who lived off the fazenda were planting on it and were subject to conditions (i.e. shares) on their production, just as the residents were. Similarly, it is very common for moradores living on Boa Ventura to have tenure on lands off the fazenda. For example, a worker who has recently migrated from a nearby plantation will continue to weed and harvest the cotton he originally planted on that plantation. It is the common policy of landlords not to purchase cotton lands from moradores who leave at will, and this may be the only way for a worker to salvage all the returns on his invested labor. Landlords fully recognize the right of an ex-morador to continue harvesting any cotton trees he has planted until they die, provided he gives the appropriate share of each harvest to the fazenda. A second group of moradores from Boa Ventura have been frustrated in their desire for good woodland to clear for their roçados, since the fazendeiro insists that this land stay fallow for the present. Because the available capoeira lands are far less fertile and much more prone to heavy weed growth than freshly cleared land, a worker naturally prefers the latter. At the time of this study, densely grown regenerated forest was available on neighboring plantations for the asking; hence several moradores, besides planting on Boa Ventura, made special arrangements to plant fields on other fazendas.

Although a worker can easily establish rights to the produce of a given piece of land, that land may at any time be appropriated to another use by the fazenda, which has only the customary obligation to pay for the labor the worker has invested. The subject of shares and "contracts" will be discussed later, but the reader should be reminded that all land the morador works within the fazenda is subject to "rent" of some sort. Essentially, the landlord, by virtue of his outright ownership, receives part of the value produced by the labor of the worker; this the landlord considers a return on his cash investment. The remainder of the produce belongs to the worker, as payment for his labor; and he may dispose of it as he wishes, though he often has few alternatives.

Residence on the fazenda, and cooperation with the farming arrangements made by the landowner and his manager, entitle each worker to a house built (and ideally, maintained) at the expense of the landlord. Upon arriving, the new worker is assigned a vacant house. If he finds it unsuitable, he may ask to move to a better house when one is available; but he must do so well in advance, since good houses are always in demand. If the fazenda manager, for reasons of economy and efficiency, should want to remove the worker to another house, he does so freely, even if the worker has made repairs on the house himself. This practice generates a good deal of grumbling (here again, the ubiquitous notion that labor should always be rewarded). Some workers who know bricklaying build their own houses; but they are always reimbursed for their labor by the fazenda, so that there will be no doubt that the fazenda owns the structure.

We have seen that a worker may often maintain continuous tenure on one piece of land for some time simply by working it. This kind of tenure may be arbitrarily ended when the landlord needs the land for something else and pays off the worker. The nature of the land and the means used to exploit it also affect the length of tenure. The rich coroas on the river margins, which are periodically regenerated by overflows from the river in heavy winters, are usually cropped continuously all year for many years in succession. Some coroas on Boa Ventura have been cultivated continuously for twenty years or more (although workers who have lived in one place for twenty years are extremely rare).

But in general, the lands available to the moradores are hillside plots, which lose fertility after two or three years and must be abandoned for five to eight years before they are cleared again. In a sense, the workers' tenure in these fields is only exercised for part of the year: only one crop a year is raised, and after the harvest cattle are grazed in the fields until the next planting season. An understood part of the tenure agreement on these lands is that the fazenda has the right to graze cattle after a given date (this date is never exactly specified, but the moradores know about when the cattle will be let in). If a morador has not completely harvested his cotton (which happens frequently, since cotton is the last crop to be harvested), he will lose it. Finally, when the cotton plants die and the land is ready for fallowing, tenure reverts to the plantation, to be allotted again when the fallow period is up.*

* This does not appear to result from shifting cultivation, as one might expect. Both Pospisil among the Kapauku Papuans (1963: 128ff) and Salisbury among the Siane (1962: 61–76) describe swidden systems in which tenure on a plot is maintained by some form of marker (*tankets* in Siane, *ti* plants in Kapauku) throughout the fallow period. That tenure does not extend through the fallow period on Boa Ventura is probably a consequence of the high migration rate, which insures that virtually no one who plants a plot will be a resident worker when the plot is available for planting again.

CROPS

Over the entire year, the moradores of Boa Ventura have available a fairly wide range of crops, but only a few of these are staples or important cash crops. Part of each year's harvest is set aside as seed for the next year; seed that has been commercially prepared is not used, nor did I encounter anyone who was aware that such seeds existed. The workers (more accurately, their wives) pick over the harvest and sort out a supply of clean, healthy-looking seed, which they store in jars until the next planting.

Table 4 lists the major crops of the fazenda. In some cases, the workers recognize many differences within the categories given. For example, the category "squashes" (*frutas*) actually contains a number of different species. Although there are names for each of these, the only important distinctions for the purposes of this discussion are those that have implications for planting behavior. Thus we need consider only one kind of squash, since all varieties are planted alike.

Other crops either planted by or available to the moradores of Boa Ventura but of less importance than those in Table 4 are:

ata (sugar apple)	lemon	peppers
avocado	lettuce	pineapple
banana	mango	plum
cashew	mint	sesame
coconut	onion	sugar cane
guava	oranges	tangerine
various herbs	papaya	tomato

Of these, the plantation raises bananas and oranges for sale in Fortaleza. Sugar cane is raised for cattle feed during the dry season; feed grasses for the cattle are also raised with the moradores' labor. Other crops are raised for consumption on Boa Ventura. A few of the above crops—e.g. onions, tomatoes, peppers, lettuce, mint, and herbs—are raised in planting trays along-

TABLE 4

Major Crops Planted on Boa Ventura

Maize (milho). Two varieties. The better-tasting *branco* is planted for home consumption. *Vermelho*, which gives more weight per unit measure and yields sooner, is planted chiefly for sale, but also provides some early corn for the workers' households.

Beans (feijão). Two varieties. Three-month beans (*de moita*) do best in low land, yield very quickly, and interfere less with manioc when the two are planted together; five-month beans (*de corda*) are suited to hill plots and give a greater yield.

Lima Beans (fava). One variety. These are always planted with maize and allowed to climb up the cornstalk. They are one of the last crops to yield.

Manioc (mandioca). Two varieties. *Manipeba* can be harvested at any time after 18 months growth, and will store in the ground for as long as six years;[a] it is used as a kind of calorie storage. *Carregadeira* can be planted in poorer soil and yields after only six months; however, it can only be harvested in July and August, and will only keep in the ground for three years. Both varieties interfere with most other crops, and they are almost always planted in the lowlands.

Sweet Manioc (macacheira). Planted like mandioca but processed differently. It is not nearly as popular as mandioca.

Rice (arroz). One variety (one morador feels there are two, and is experimenting to find just what the differences are).

Potatoes (batata). One variety of sweet potato.

Squashes (frutas). This is really a group of crops, including squashes, cucumbers, pumpkins, watermelons, and gourds. Each type is consumed differently, but all are planted and harvested in the same way. The workers themselves apply the term frutas to the group as a whole.

Cotton (algodão). Two varieties. *Mocó*, a tree cotton usually planted in hill plots, yields poorly the first year; but it then yields for 8 to 20 years before dying. *Herbáceo* is a low-growing variety that yields very well but dies after one year; it grows best in humid, low-lying soils, especially under irrigation.

---

[a] Jones (1959:22) reports that manioc roots become woody and difficult to process as they age, so the six-year figure given by one of my informants may be excessive.

---

side the houses by the women, who otherwise have little to do with agriculture. Except for medicinal herbs, there is little or no use of wild plants.

Crops and land. As we have seen, the workers categorize both the crops they sow and the lands they sow them in. However, they do not do this arbitrarily or in isolation from other facts. They have an entire complex of ideas and techniques concerned

TABLE 5
Classification of Lands and Crops

Land Typology

Strong

Hot

ROÇADO

COROA

RIO
With fertilizer

LAGOA, BAIXO

Cold

CAPOEIRA
2d Year

CAPOEIRA
3d Year

RIO
Without fertilizer

USELESS
Saline

USELESS
Dry, Sandy

Weak

Observed Planting Pattern

Strong

Hot

SQUASH
Roçado only

MANIOC,
3-MONTH BEANS
Mainly Coroa,
occasionally Roçado

RICE
Roçado, Coroa
Baixo, Lagoa

HERBÁCEO COTTON
Coroa, Lagoa

5-MONTH
BEANS
Roçado,
2d-year Capoeira

MAIZE
Roçado,
2d-year Capoeira,
Coroa

Cold

MOCÓ COTTON
Roçado,
2d-year Capoeira,
3d-year Capoeira

POTATOES
Mainly Rio, with
or without fertilizer

Weak

with getting the highest yields from available materials; crop and land distinctions are just some of the conceptual weapons that help in the workers' struggle with nature.

The distinctions between types of land involve two basic qualities, each a continuum between two extremes. The first is the continuum from "hot" to "cold," a rather complex and variable evaluation based on the moisture content of the soil (see pp. 43–44). The second quality is a continuum from "strong" to "weak"; this is a straightforward evaluation of a field's capacity to give a high crop yield, in any given year and for many years running. Each type of land is distinguished from the others by its positions on these continua.

Crops may be distinguished from each other by the planting behavior of the workers, which closely reflects this scheme. Table 5 represents fairly well the types of land recognized by the Boa Ventura workers and the major crops usually planted in each. One point should be made: workers often mentioned that land was "too weak," "too hot," or "too cold" for a given crop; but none said that land could be "too strong." Therefore, I have listed each crop at the lowest approximate level on the strong–weak continuum at which it may be successfully planted. A crop may be planted at any higher level and do well, but not vice versa. For example, maize is planted in either hot or cold land, but not in the riverbed nor in a capoeira after the second year; and because it is a staple crop, it is almost always planted in the stronger coroas and roçados when possible. Squash is planted in roçado (especially in the rich ash areas of burned-out tree stumps), and only rarely in the lower, colder areas. The scheme is important for what it lacks as well as what it includes. Note that there are no distinctions based on whether the soil is plowed, whether the crops are fertilized (the river bottom, planted in the dry season exclusively with potatoes, is the unique exception), or whether pesticides and special seeds were used.

Of course, this table is artificial and merely expresses tendencies; in practice, many exceptions occur. For example, a worker may plant potatoes in the riverbed without applying manure; he knows he will get less than half the yield, but he may have neither the time nor the equipment necessary to transport manure from the stable. Or a worker, knowing that three-month beans do best in cold land, may nevertheless plant them in hot land near his house, so that, should the river rise and bar him from his cold lands on the other side, he can still harvest a few beans early in the rainy season. In sum, planting decisions are definitely influenced, but not exclusively determined, by the concepts summarized in Table 5.

In the course of my fieldwork, I collected detailed information on the economic activities of 44 heads of households. From this data, I hope to give the reader a quantitative picture of planting and yields that may easily be compared with more familiar agricultural systems.

To begin at the most general level, it might be useful to describe a sort of average behavior. Of the numerous types of land distinguished by the workers, five are especially important: roçado (swidden, first year), capoeira (swidden, second and later years), coroa (fertile river margins), *rio* (the dry riverbed, used only for potatoes), and *baixo* (moist, low-lying areas scattered in pockets around the fazenda). In addition, crops are sometimes planted in *lagoa*, the moist land on the edge of a receding reservoir. The moradores have varying access to these different types of land, and many plant in more than one kind. Fully 74 per cent plant in roçado, which may be considered the standard form of field; 57 per cent plant in capoeira, either as their main field or to supplement a roçado; 34 per cent are able to plant in coroa, although few can do so in amounts large enough to support a family without some other field as well; 18 per cent plant potatoes in the rio during the dry season; and only 11 per cent of the workers plant crops in baixo.

By taking simple arithmetic means, I calculated the average configurations of planting in each different type of field ("average" because, as we shall see, individual differences are great). A comparison of the major crops planted in roçado and capoeira

TABLE 6
Average Mixed Planting Behavior, Roçado and Capoeira

Crop	Seed Planted in Roçado of 2.8 Hectares	Seed Planted in Capoeira of 2.8 Hectares
Maize	23.2 liters	12.8 liters
Beans	5.8 liters	4.0 liters
Cotton	6.2 kilos	2.4 kilos
Lima beans	2.2 liters	2.8 liters
Rice	1.3 liters	0.6 liters
Squash	7 plants	3 plants
Potatoes	few	few

is particularly striking. Both were planted with a mixed crop; but on the average, a second-year field was planted with only about one-half of the seed used in its first year, so great was the expected drop in fertility (see Table 6). The figures do not show, however, that in the second year most workers planted only maize and beans, whereas during the first year nearly every field had the whole range of crops, as well as some bananas, and sesame. The cotton plants (mocó, or tree cotton) survive for many years, so cotton planted in the second year is usually a replacement for plants that died during the first year. Table 6, of course, shows a typical mixed cropping; on occasion, the moradores plant a one-crop field of manioc or potatoes in a hillside plot. These two crops, according to the workers, interfere with most of the others in one way or another, so they are not planted in any numbers in the usual mixed-crop field.

Coroa land, which is very desirable and in short supply, is normally held in plots averaging 0.9 hectares, about one-third the size of an average roçado or capoeira (2.8 hectares); in the amount of maize planted, 7.7 liters, they are proportionately more similar to roçados than to capoeiras. Maize, beans (three-month variety), squash, bananas, and sesame are usually present; but coroas, unlike roçados, have only small plantings of lima beans, rice, and cotton. The most striking distinction of the coroa, however, is that the average field contains over 3,000 manioc plants, and coroa thus provides most of this very important food.* The riverbed is planted only in sweet potatoes, the average plot holding 1,600 plants. The humid baixo and lagoa land is of relatively little importance, and is reserved for small fields of rice, or in one case for a crop of 2,000 manioc plants.

So far I have been talking as if differences in land type were the only determinants of differences in the moradores' planting behavior, but this is not the case. If they were free to choose, they would plant at least a few crops in each of the land types, in order to minimize their potential loss in a sudden wet or dry spell. But the various land types are very unevenly distributed around the fazenda, and where one lives does much to determine the kinds of land one may plant crops in.

In this regard, there are three significantly different sections of the plantation (see Figure 3). The 44 households represented in my economic census are distributed as follows: 16 are just northeast of the river, with access to a wide variety of lands; 10 are in the extreme northeast, which offers only hillside swidden land in a sparsely populated portion of the plantation, distant from a water supply but near abundant regenerated woodland for clearing new roçados; and 18 households are southwest of the river, close to water and shops but in an overpopulated area where land for clearing is so scarce that many farmers are not able to plant new roçados. Table 7 shows the effects of these sectional differences.

I also found substantial differences in the area planted by each worker, and in the proportions of the various crops he chose to plant. For example, several households planted a total of less than one hectare for the year, whereas two others planted eight and five hectares, respectively (using mostly hired labor).

* Manioc cannot be conveniently planted in roçados, not only because it interferes with other crops but also because cattle will eat the leaves of the plant after the cotton harvest, before the manioc root is ready for harvesting. In any case, the moist, fertile coroa is probably better for manioc than the hillsides. Although manioc can grow in areas receiving anywhere from 20 to 200 inches of rainfall a year, and can sustain long droughts (hence its great value in Ceará), it grows best in soils that are "moist, fertile, and deep" (Jones 1959: 16).

TABLE 7
Number of Fields Planted, by Type

| Section | Houses | Number of Fields | | | | | Total Fields | Number of Fields per Householder |
		Roçado	Capoeira	Rio	Baixo	Coroa		
One	16	13	9	8	3	10	43	2.7
Two	10	10	3	0	1	1	15	1.5
Three	18	9	13	0	1	4	27	1.5

Furthermore, if one takes the amount of maize seed planted as a base,* then some men planted relatively more beans per unit of maize, or less squash, and so on, than others did. These differences can only be considered idiosyncratic, depending ultimately on the abilities and aims of the individual worker, the size of his household labor force, his beliefs regarding the value of hiring labor or of using particular crops, etc. However, each worker seems to have these idiosyncrasies, and each worker's total planting behavior has a configuration of its own.

Crop yields. With luck, and proper technique, the morador on Boa Ventura can receive very good returns from his fields. He is usually able to plant in fertile soil, and gives his fields close attention as the crops develop. But the morador has no control over the weather, and the harvest varies greatly in this unstable climate. By my informants' standards, 1965 was a moderately good year, with an adequate, evenly spaced rainfall. By contrast, 1966 was a bad year—not a full-fledged drought, but dry enough that many plants never matured and others yielded poorly. The figures in Table 8 for yields of maize, beans, and rice show how these crops suffered from the dryness, even though they were grown in the same kinds of fields and with the same techniques.

These three important crops, particularly rice, are quite sensitive to rainfall. By contrast, manioc and cotton, which are also very important in the workers' economy, did not show any sig-

---

* The amount of maize planted in a field, rather than some area measure, is the workers' means of expressing the size of a roçado. Usually, about 8.3 liters of maize are planted per hectare of roçado.

nificant fluctuations from 1965 to 1966. Manioc, as I have noted, is planted in large stands along the river; it matures over a period of several years, although it may be harvested at any time after the first year or so. On the average, workers end up with about 400 liters of manioc flour for every thousand plants under cultivation. But this figure fluctuates wildly from worker to worker owing not to the weather but apparently to the amount of care the plants have received and the amount of time they have stayed in the ground. Thus reported yields run from 80 liters to about 2,000 liters per thousand plants, although the overall totals are fairly constant. Cotton yields are markedly consistent from year to year and field to field (except during the first year, when the immature plants give a very low yield), averaging 183 kilograms of cotton per hectare.

Each kind of land has its own influence on crop yields (see Tables 8 and 9). For example, coroa, which is much like roçado in fertility, has a steady water supply from the river and therefore varies much less from rainy to dry years. This quality, more than any other, is what makes workers eager to plant at least

TABLE 8
Average Yields (All Sections, Roçado)[a]

	Maize			Beans	Rice
		Yield	Yield per	(Yield	(Yield
		Yield	Unit Seed	per Unit	per Unit
Year		per Hectare	Planted	Seed)	Seed)
1965 (wet)		1,180 liters	130 : 1	148 : 1	92 : 1
1966 (dry)		819 liters	96 : 1	66 : 1	12 : 1

[a] For average amounts planted, see Table 6, p. 59.

TABLE 9
Maize Yield, Capoeira and Coroa

	Yield per Liter of Seed		Liters Harvested per Hectare	
Year	Capoeira	Coroa	Capoeira	Coroa
1965 (wet)	75 : 1	118 : 1	635	909
1966 (dry)	48 : 1	116 : 1	382	1,091

TABLE 10

Plantation Total Sown and Yielded, 1967

Crop	Roçado		Capoeira		Coroa	
	Sown	Yield	Sown	Yield	Sown	Yield
Maize (liters)	741	96,330	321	24,075	116	13,688
Beans (liters)	185	24,380	100	10,000	18	900
Lima beans (liters)ª	70	4,690	70	4,690	1.5	100
Rice (liters)	41.5	3,818	15	750	3	300
Squash (no. of fruits)	230	1,150	75	375	18	90
Manioc (plants/liters)ᵇ					47,220	18,888
Potatoes (plants/kilos)ᶜ					2,700	932

ª The figures for lima beans are approximate, but close to the truth. Surprisingly, at least as many limas are planted in capoeira as in roçado, which is not the case with any other crop.

ᵇ Manioc is propagated by cuttings, and I have given the number of plants set out. The harvest is almost all taken from plants sown in previous years, and does not relate to the manioc planted in this season.

ᶜ A very few potatoes were planted in roçado plots.

a small plot in coroa. It is also clear from Table 9 how much less fertile than either roçado or coroa are the capoeira fields. After the second year, capoeira is devoted entirely to the cultivation of the fully mature cotton plants; so the maize was planted in the best capoeira, only one year removed from roçado. We can see in quantitative terms that the decision to stop planting food crops in a hill swidden after the second year is based on a very real decline in land productivity. It is not clear whether this drop in fertility is due to a loss of nutrients in the soil or to competition from the increased growth of weeds (cf. Carneiro 1961: 55–57).

Unfortunately, I left the fazenda well before the 1967 harvest was fully under way. However, I did obtain data on 44 households recording what amounts of each crop they planted in what kinds of land. Since 1967 was a year of good weather like 1965, I have used the average yields calculated from 1965 as a base from which to predict the total harvest for 1967 (see Table 10). When these figures are compared with my data for labor input during the 1967 growing season, it will be possible to measure household labor productivity and household consumption.

Besides the crops planted in the major fields, 17,000 potato plants were planted in the rio plots, which would yield an estimated 5,865 kilograms of sweet potatoes at harvest. In the humid baixos, 13 liters of rice and 2,000 manioc plants were sown, which would produce 1,534 liters of rice and 800 liters of manioc flour. Finally, the average household usually harvests 1,090 kilograms of cotton from its various fields; the total fazenda yield (for the 44 households in the survey) would thus be 47,960 kilograms of cotton.

AGRICULTURAL TECHNIQUES

Planting methods. In describing planting, I shall be concerned mainly with the manner in which moradores calculate their planting ahead of time and spread their efforts over a range of micro-environments in order to profit in almost any weather. But first, a brief mention of the actual techniques of planting is in order.

In Ceará it is impossible to predict when the first rains will come. Winter is preceded by light showers; these are of no value for crops but do germinate the grasses that will feed the cattle over the next six months. Even in years of normal rainfall, winter may begin as early as December, or as late as March. The first good, groundsoaking rain of the year is the signal to begin planting. Because the first rain is often followed by a long dry stretch, however, few workers put their eggs in one basket by planting all at once; rather, they plant in the land they know to be dampest, and adopt a wait-and-see attitude about the rest. When a second good rainfall comes, they feel secure enough to go out and plant all their remaining lands.

Planting, like most agricultural labor on Boa Ventura, is done with a hoe. It takes at least two workers: one to open the ground with the hoe, and another to drop in seed and cover it over with dirt. More than two are preferable, since a fast hoe worker can keep two or three others busy dropping the seeds. The field is "aligned," as informants put it, by planting the maize first (and with it lima beans, if they are planted), for the rows of maize form the framework within which other crops are sown.

Mixed cropping is the rule here, as in many swidden agriculture systems (Conklin 1957, Geertz 1963: 16–19).* It sometimes happens that maize, two kinds of beans, lima beans, cotton, rice, squash, and potatoes will be sown together within the same field. Such crowding is unusual, but some mixing always occurs in roçados and coroas. Second-year clearings are usually limited to maize, beans, and cotton. Figure 5 illustrates some of the variations I observed in the distribution of crops in the hillside roçados. Depending on what crops are planted, various schema are selected, all tending to space the crops so as to reduce their interference with one another. Beans and lima beans need room to spread and grow. However, lima beans are always planted together with maize because they need something to climb; since they grow slowly and yield months after the maize, they do not interfere with the maize plant appreciably. But differences in planting behavior do not always represent a straightforward adaptation to ecological features. A good deal of the variation results from the inability of the moradores to agree among themselves on the best methods. I wish to emphasize this point. Anthropologists have too often reported primitive or peasant agricultural practices as though each individual farmer followed the same set of rules based on the same cultural understandings. This is certainly not true of the Boa Ventura workers.

An example, trivial in itself but representative of a whole host of others that could be given, is the decision of whether to plant beans in rows between the rows of maize, or to plant them together with maize in the same hole. Out of 22 workers I questioned on this subject, 11 stated that they had tried both methods and were convinced that planting beans between the rows of maize gave a much higher yield than the other method; hence they employed this method, even though it entailed the extra labor of cutting entire new rows of holes with the hoe. (Even

* Geertz (1963: 16–24) enumerates the ways in which swidden systems simulate tropical forests, and the system found on Boa Ventura is no exception. First, a great variety of species is planted in one plot. Second, burning the field returns nutrients stored in the natural vegetation to the soil, for later storage in crops. Third, crops are planted "in a tightly woven, dense botanical fabric."

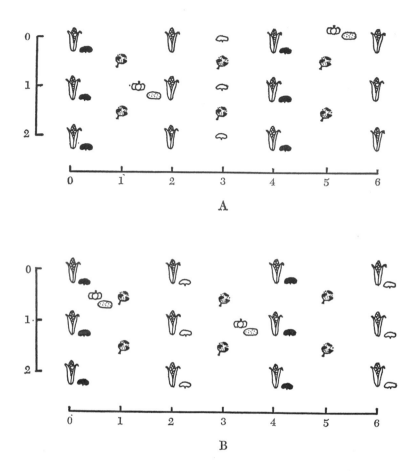

Figure 5. Roçado planting patterns (measurements are in meters).
Sketch A shows a small section of the most common type of hillside roçado.
The crops are well spaced, with five-month beans and lima beans in alternating distribution. The random distribution of the few squash and potatoes
planted is determined by the location of burned-out tree stumps. Pattern B
employs the labor-saving device of planting both kinds of beans next to the
maize plants, eliminating the need for an extra set of seeding cuts with the
hoe. In pattern C rice has been added; and all beans are planted in the
same furrows as maize. Pattern D shows another method of planting rice,
which allows an unusually varied crop mix. It was planted by one of the
brighter young workers, and is even more complicated than it looks because
both three-month and five-month beans were planted.

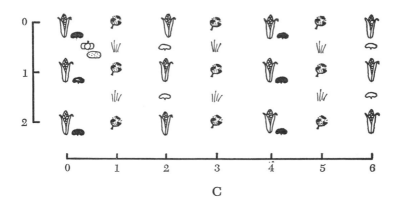

C

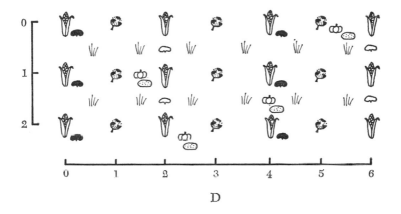

D

LEGEND

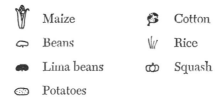

Maize   Cotton

Beans   Rice

Lima beans   Squash

Potatoes

here there was disagreement: some workers insisted that the additional labor was equivalent to the labor already invested in planting maize, but others felt it was only one-third that amount). Another five of the moradores agreed that the first method gave higher yields of beans; but they either could not spare the labor or felt that the added returns did not warrant the extra labor, and so planted beans simultaneously with the maize.

In this case, 16 out of 22 workers agreed on the basic parameters of the problem, but made different decisions on the basis of their own personal evaluation. Another six did not even agree that beans planted between maize yielded more—their own experience had not taught them that there was any significant difference in yield, and they selected, understandably, the method of least labor.

A few other examples should make clearer just what is going on. All the workers agree that potatoes planted in the riverbed after the river dries up will yield much more if manure is carted in from the stable and used to fertilize the potato plants; but some moradores are too poor to spare the time, or to get hold of a donkey to cart the manure, so they accept a lower yield. Only one worker insisted that there were no differences in yield between the two methods. Again, some workers will not plant rice at all because they feel the risk is too great; others take the risk. Some are fond of lima beans and plant them; others dislike them and refuse to plant them. One man dislikes them as food, but plants them in order to trade them later for a preferred food. It should be clear that there are two broad kinds of disagreement present. In one case, the workers agree on the most desirable alternative but make different decisions owing to unequal capabilities (e.g. possession of a donkey). In the other case, the workers disagree at the most basic level on how their natural world is structured (e.g. rice is risky). Each case accounts for some of the variation in decision making.

Obviously, the common understandings that we like to think make up a cultural system are not universally shared by these

Sujeição labor. The workers are clearing brush from a roçado belonging to the fazenda manager. The larger trees will be chopped down later.

A newly burned field. Some stumps are still smoldering; later, potatoes and squash will be planted in their ashes. This field belongs to the manager and is well above average size.

A small fazenda in the sertão, some distance from Boa Ventura. The gardens around the owner's house are the only irrigated land. The house itself is typical of those in the area.

The blacksmith of Boa Ventura at his forge. He derives almost all his income from his trade, and is the nearest to a full-time specialist on the fazenda.

Sujeição labor. Workers are hoeing weeds in one of the landlord's sitios. These lowland fruit gardens, being irrigated, are the choicest plots on the fazenda.

Woman by the fence of a sitio.

Workers clearing weeds from the edge of an irrigation ditch, which carries water from a large public reservoir to a number of local users. The landowners usually make jobs like this available in drought years to tide the moradores over.

(*Below*) This ten-year-old boy can earn about Cr$500 a day (18 cents) by loading bricks and moving them to construction sites on the fazenda.

(*Above*) The vaqueiro's son, mounted on one of his father's horses. He helps with many of the routine herding duties on the fazenda.

The vaqueiro and his son driving the fazenda cattle to pasture during the dry season.

The landlord's truck loaded with cotton for transport to the city of Fortaleza. This truck was the present landlord's major investment in the fazenda.

The manager's son scraping bristles from a pig that has just been slaughtered. A few cuts of the meat will be distributed as gifts to workers who helped with the slaughtering.

Harvesting potatoes in the empty riverbed late in the dry season. The man at the left has just returned from hunting, and carries his rifle and a sack for small birds.

An old woman grinding coffee beans in the traditional manner.

A worker and his family standing in front of their home. The building is that shown earlier with the landlord's truck parked in front, and is near the center of the fazenda.

The manager of Boa Ventura standing in front of his house with his third wife on his right, several of his 35 children, and a few guests. All are wearing their best clothes.

moradores. Nonetheless, there are a few environmental characteristics crucial enough in their effects on planting to lead to general agreement among the moradores. One such is the difference between hillside and lowland fields.

In the lowlands, the hillside practice of alternating crops with cattle during the year is not followed because the land is moist ("cold") enough to permit either manioc raising or the sowing of two crops a year. These two possibilities have important consequences. First, since the workers never plant manioc together with cotton or five-month beans, coroa fields of manioc show much less crop mix than hillside plots. Second, since there is the possibility of two crops a year where manioc is not planted, the moradores plant their fastest growing varieties in order to harvest twice before the rainy season has completely ended. It is in the low-lying fields that farmers sometimes plant a one-crop lastro, usually in beans; and only there can one plant the important fruit trees, such as bananas, papayas, mangoes, and cashews. In some low baixos, water is standing a good part of the year; these are reserved for rice, which is often started in a small garden and then transplanted to the wet field.

One major purpose of my discussion so far has been to show that the Boa Ventura workers are required to make a great many decisions throughout the year; and that these decisions are not guided by easy cultural formulas, but must be made in accordance with the particular understanding and circumstances of any given morador. The workers are not, it seems to me, in any way careless in making these decisions, at least in the sphere of agricultural production. They plan as much as a year in advance (more in a few cases), in order to save labor by cooperating efficiently with their neighbors and to use all the resources at their command effectively.

I can best illustrate this by describing the care with which one of the more thoughtful—and successful—moradores made some of his agricultural decisions in 1966 and 1967. This worker planted in seven different fields, representing six different environments: a hillside roçado; two coroa fields; a secondary river channel, which was filled with water only in years of heavy rainfall; a wet lowland; the edge of a reservoir; and the riverbed after it had dried up for the summer.

He had asked permission to clear the hillside plot at least six months in advance of planting season, so as to be able to begin preparatory chores during the previous year's cotton harvest. In allocating his labor, he was especially careful to finish clearing his land before the date on which all the fields in the area were to be burned off for planting. In this roçado he planted maize, beans (including some of the three-month variety to provide an early harvest of food if his others should fail), cotton, potatoes, squashes (including squash, pumpkin, watermelon, and gourds), and lima beans. This worker had also established an exclusive right to continuously farm two small fields on the river margin. Here he planted maize, three-month beans and manioc.* Short of a drought year, or one so wet that his hillside crops would be carried away in landslides, he would be assured of at least some crops from one of the two plantings.

In a dry secondary river channel, rich with floodwater deposits, he planted banana trees, intercropped with maize. He had noticed that this land, which could easily be watered from a nearby irrigation ditch, was unused, so he cleared and planted it. During my study, he was experimenting with a new method of planting banana trees, which was then being pioneered on a neighboring, technically advanced plantation; he planted half of his trees in the new fashion. and left the others as a control. Since the river could easily overflow into this channel if there were heavy rains, he had planned his planting very carefully in advance. Immediately after the first rain, he planted the maize (among the now full-grown banana trees). If the maize grew fast enough, he could run down and pick the green ears when the river began to overflow, salvaging at least some of the crop. If the river overflowed too soon to allow this, he

---

* He planted both early ripening manioc (for picking after one year) and late manioc, in order to have as long a time span as possible in which to pick manioc—a kind of insurance against bad years.

would have time to plant a second crop; this had happened in the previous year, when he only harvested a crop after the third planting.

In the case of the wet lowland, this worker again saw the unused potential of a fallow piece of land—a low spot near a dam, into which the overflow from the dam spilled at the peak of the winter rains. He planned to plant rice there when the first rains fell, realizing that by the time the dam overflowed the rice would have grown tall enough to require standing water. When the river had dried he planted 2,000 potato plants there, having arranged at least six months in advance to use someone else's donkey for hauling manure to the field. Finally, along the edge of the reservoir he planted a small crop of beans. These vazante areas are marginal and of poor quality in a good winter. But should there be a drought, the reservoir would shrink, exposing more and richer land—the only land wet enough for planting. The drier the year, the more productive is this land, and the more critical its importance.

This planting scheme, though more extensive than many workers can manage, is fairly typical. Perhaps its most interesting aspect is the "hedging" strategy used for planting—investing labor in a wide range of alternatives to reduce the risk that any single "unmixed" strategy is exposed to. This is a particular case of a general rule followed by the moradores in making a great many decisions: to prefer a strategy that offers security to one that contains risk, even though the latter offers the prospect of greater immediate returns. Workers will try to second-guess nature on occasion, planting a crop that might fail even under average rainfall conditions. But this gambling is limited to a small fraction of the total agricultural investment of most workers; almost all their energy is directed at assuring a secure food supply for the following year. One effect of this cautious strategy is that just about all the kinds of land available on Boa Ventura are being exploited. (This is not to say that all the workers have an equal variety of land under cultivation; after all, some land types are in relatively shorter supply than others.)

Aside from hedging, the moradores try to increase their overall security by using two further techniques employed by the
worker I have described. First, they always plant a few quick-
yielding crops of lower productivity, so that the period of food
shortage immediately before the harvest will be as short as
possible. Second, they try to plant at least some crops relatively
less sensitive to rainfall fluctuations and productive over a long
term (e.g. cotton and manioc); these will ensure future earnings
even in years of bad weather.

Farming technology. The tool kit employed by the moradores in their subsistence activities is extremely limited, usually including only hoe, ax, and billhooks. One might conclude
from this that the agricultural technology of Boa Ventura is
"backward." This opinion implies that more "modern" techniques exist, and that the workers (or any other peasants)
should, by some objective standard, be using these. Two reasons are often given to explain their backwardness in not doing so.

First, it is held that peasants are bound by tradition to pursue old, established ways of doing things. In its most insulting
form—that peasants simply cannot see beyond their traditions
to obviously superior methods—this notion is rapidly disappear-
ing from the literature. But it is being replaced by an essentially
similar, though less condescending, notion: that peasants have
learned through bitter experience to follow certain paths, which
they are understandably loath to change, considering the price
paid to learn them (Rogers 1969: 31–32). This idea contains
both an element of truth and an important error. The truth is
that peasants have usually acquired their present techniques
through experience, testing various techniques and adopting
the successful ones; the error is in assuming that this peasant
pragmatism has anything to do with conservatism.

In the history of Ceará, many major innovations—such as
steel tools, new varieties of crops, and the use of chemical fertilizers—have been accepted quite rapidly, whereas others,
such as the plow, have spread slowly or not at all. Apparently,

the "negative cultural conditioning" that supposedly retards innovation is operating selectively. Actually, it is not negative conditioning that is operating here; rather, the peasant, at least on Boa Ventura, is an active pragmatist who is chiefly concerned with whether or not something works to his own material advantage (cf. Schulz 1965: 163–68). That moradores do experiment, as documented in these pages, is one of the strongest arguments in favor of this view.

A second and more sophisticated reason is often given for peasant backwardness: the peasant cannot innovate because he is too poor to do so; that is, he simply does not have the capital to acquire plow and team, or to install the motors, pumps, and pipes of an irrigation system. The low capital and savings of the average morador are clearly an important fact of life on Boa Ventura; and the great emphasis the workers place on security is clearly a result of their lack of personal wealth, which would cushion them from chance disasters. But the workers' poverty alone cannot entirely explain the simple, spare tool kit of the backland farmer.

The tool kit of the morador is a typical one for swidden agriculture, and it represents but a small part of swidden technology. This technology itself is often considered a sure sign of backwardness; however, it is well established from comparative studies that swidden farming, below a certain level of population, is an efficient and productive technology under tropical forest conditions (Conklin 1957, Geertz 1963, Meggers 1957). "Improvements" in this technology may lead to no real improvement in the lives of the workers. This is the important consideration, for we cannot expect a pragmatic worker to accept an innovation that does not perceptibly improve his lot. Boserup (1965: 65–69) has argued convincingly that the movement from extensive to intensive land-use systems has been in response to increasing population, and not vice versa. Each step that increases the productivity per land unit requires a disproportionately greater labor input.

It is clear from the evidence presented in this chapter that present labor inputs give the Boa Ventura workers an adequate food supply in most years. Further, the mild labor shortage in the region indicates that there is no great population pressure on land resources at the moment. Given these two conditions, we would actually expect a peasant to reject more laborious techniques unless he was sure of increasing his total income in proportion to his extra effort.* I would argue, then, that the limited tool kit and technology of the morador are adequate under present population pressures; and that we cannot reasonably expect the moradores to adopt such labor intensive techniques as pesticides, chemical fertilizers, and complex irrigation until land shortage forces him to. With these considerations in mind, let us turn to the technology employed on Boa Ventura.

The morador's most important tool is the hoe: from the beginning of planting until well beyond the harvest, it is practically the only tool the farmer uses; with it, he plants seed, and weeds his field periodically throughout the growing season. All adult males, and most males over ten years old, have their own hoes. In addition, most households have a few brush-cutting tools (chiefly billhooks and machetes) for use in the initial clearing of fields and in keeping the old cotton fields free of brush. Finally, each household has one ax, which is used to fell the large trees that remain in the roçado after the brush has been cut away. With very few exceptions, the moradores on Boa Ventura have no other tools. A few men own picks, which are used to dig postholes for fences; three men on the fazenda have carpenter's or mason's tools; and finally, one worker owns a cultivator, pulled by a team of bulls,† and borrows a plow from a friend at planting time.

As soon as I found that only one man used traction animals,

---

* Boserup writes (1965: 66): "Reports of extension officers . . . give numerous examples of cultivators who refuse to introduce ploughing or transplanting or production of fodder or other changes suggested by the advisers with the explicit motivation that this would add too much to the labour with the crop. Such objections are often interpreted as a lack of interest in raising total income, but it may be suggested that they can be more plausibly explained as the result of a quite rational comparison between the additional labour and the probable addition to output."

† Bullocks are not castrated before harnessing for traction.

and that he was a newcomer to the plantation, I asked workers how they felt about his farming techniques. They agreed that a cultivator drawn by a team of bulls was far superior to a hoe, clearing weeds more thoroughly at much less expense of human energy. Here were peasants who knew the value of a tool and thought it very desirable; but for some reason there was only one who actually used it, and he had acquired the skill elsewhere. The moradores gave various reasons for this situation. For one thing, a cultivator requires an initial investment that is not entirely out of reach but certainly makes a worker think very hard before spending the money. Moreover, the use of such a tool requires at least one and preferably two animals. This difficulty is almost impossible to overcome, for bulls are very expensive; besides, the fazendeiro allows practically no one to own cattle, horses, or mules, since these compete for pasture with the fazenda cattle. A final, and very important, reason is that the hilly uplands where most workers raise their crops have slopes too steep for a team to work on; hence only a fraction of the workers could use a team in any case.

Manuring and irrigation are two techniques in widespread use on the fazenda but their distribution is very uneven. Irrigation can only be employed in the flat lowlands; irrigated, the fields here become *sitios*, which produce fruits and vegetables for market (usually, these crops belong to the landlord). All nine cases of irrigation by workers on Boa Ventura occur in one or the other of the two areas that have sitios. Similarly, all but three of the 29 moradores who use manure in their fields live in households close to the stables. In both these cases, a worker's distance from the source of necessary supplies helps determine whether or not a given technique is used. By way of contrast, the use of insecticides is also widespread on the plantation, and does not show this unevenness of distribution. The most common poison, which is used in very small amounts on leaf-cutting ants, is cheap, highly portable, and accessible to everyone. Slightly less than half of the moradores actually do use it; the others feel that it does not increase harvests enough to be worth even a small cash outlay.

Steel tools and chemical insecticides are the only industrial products generally used by moradores. However, they are constantly experimenting with new kinds of crops in the hope of discovering one that yields more or is more resistant to drought and disease. As one morador explained, "Whenever we hear of something new, we like to try it out." At the time, he was experimenting with a new kind of manioc that was said to yield sooner than any other. Another worker was experimenting with a new way to plant bananas, using alternate rows planted in the traditional way as controls; and still another had just planted the seed of a giant cucumber he had acquired from a distant plantation. When I found that none of the workers had heard of hybrid seed, I explained its great advantages over the uneven quality of seed normally used. They immediately asked me to get them some so that they could try it out, and I had to admit to them that hybrid seed simply could not be purchased in Brazil's Northeast.

Usually, planting experiments are conducted on a small scale so that the risk is small. A few workers are bolder, but they run great risks. Antonio M., for example, planted several thousand tomato plants in 1965, cared for them, and received a bountiful harvest. But he lost most of the harvest because he could not find a truck to get his crops to market before they spoiled. Antonio, like other workers, learned not to invest too much in an uncertain enterprise.

I have gone into this in some length because I feel it necessary to overcome the widespread view of the peasant backlander as conservative and suspicious of change. The backlanders of Ceará may be limited in their access to information, but they are nonetheless receptive to new ideas—a situation that the development agencies of the national government have not even begun to exploit.

LABOR AND RETURNS

Labor inputs. For my purposes here, a man-day is defined as the amount of farm labor a *married* male performs in one day (an average day involves 6–8 hours, with a certain seasonal variation). The estimation of labor inputs on Boa Ventura is an extremely difficult task because there are so many differences between the persons who actually appear in the fields to work. Some differences I have simply ignored: men differ in strength, age, and perseverance, and the very tasks they perform demand different efforts and skills; there are no measures of these differences in the account given here. But other differences are important enough that some allowance must be made for them. If we accepted the above definition without qualification, then a considerable portion of the actual labor force of the fazenda would be ignored. In particular, I wish to take account of single young men and women, and even of young boys. Small girls and their mothers contribute so little to the labor in the fields that they may be ignored without affecting the results.

An unmarried male of more than 15 is fully capable of performing a man-day of labor, and may do so on the days that he works. But on Boa Ventura youths generally avoid work when they can get away with it, and on any given day one may see small groups of them standing around talking, or occasionally playing cards. Nonetheless, although his motivation is certainly less than that of a married man with family responsibilities, a grown son is definitely expected to contribute to the household effort; and the more obedient sons may approach their fathers in a year's labor output. I have placed a young bachelor's output at one-half of a married man's for the same period of time, which seems fairly close to reality.

A younger son (10–14 years old) is also allowed, and even expected, to contribute to field labor; but he is weaker than the older men, so he does considerably less work in a day, and there are certain tasks that he cannot perform at all (e.g. cutting down trees or dense brush). He is often needed at home to carry water or perform other tasks for his mother, and so contributes nothing to the crops. And in general, he is able to shirk where his older brother would not be able to—i.e. he plays more, goes swimming, or simply stays home. A young boy's contribution in the course of the year may be set at one-fifth of the standard unit.

TABLE 11
Labor Requirements on Boa Ventura, 1967

Type of Field	Task	Man-Days per Average Plot	Estimated Man-Days for Whole Fazenda
Roçado	Prepare	48.7	1,487
(average size	Plant	7.0	227
2.8 hectares)	Weed	53.4	1,737
	Harvest	34.8	1,133
	Total	143.9	4,584
Capoeira	Prepare	9.0	229
(average size	Plant	3.8	98
2.8 hectares)	Weed	32.0	818
	Harvest	19.2	491
	Total	64.0	1,636
Coroa	Prepare	5.4	83
(average size	Plant	4.6	71
0.9 hectares)	Weed	24.6	378
	Harvest	11.6	177
	Total	46.2	709
Cotton Capoeira	Weed	26[a]	1,144
(average size	Harvest	91[a]	3,995
2.8 hectares)	Total	117[a]	5,139
Total labor requirement for 1967			12,068

[a] Man-days per household.

Not all girls over 15 contribute equally. Those who live in households without older sons may be involved at all levels of the production process except the heavier tasks. Almost universally, they make some contribution during the cotton harvest, and I have seen girls at work planting and weeding with hoes. Usually, however, their labor is directed by the mother of the family, and they rarely work in the fields. Although they are indispensable during the cotton and some other harvests, their aggregate contribution is only about one-tenth of the standard in the course of a year.

Using these rough estimates, it is a straightforward matter to calculate from my household census for 1967 the total labor output of the fazenda in man-days. There are: 49 married men (49 units); 29 unmarried males over 15 ($\times$ 0.5 = 14.5 units); 22 boys aged 10–14 ($\times$ 0.2 = 4.4 units); and 22 unmarried girls

TABLE 12

Allocation of Labor (Plantation Total, Estimated from Extensive Survey)

Activity	Man-Days	Per Cent of Total
Labor in workers' own fields	12,068	47%
Fazenda obligations	1,768	7
Wage labor	2,878	11
Sundays and holidays	3,855	15
Residual (illness, idleness, etc.)	5,017	20
Total	25,586	100%

over 15 ($\times$ 0.1 = 2.2 units). The total comes to 70.1 units, or man-days, available each day. Hence there were about 25,586 man-days of labor potentially available on Boa Ventura in 1967.

I also collected more detailed statements from 32 heads of household regarding their average labor inputs in specific agricultural tasks. This information is summarized in Table 11. I have broken down labor by task and type of field, expressed it for an average size of field and estimated it for the plantation as a whole.

Now, let us see how the total agricultural man-days available on the fazenda are allocated (see Table 12). First, 52 of the available days in 1967 were Sundays, on which no agricultural work is performed (the only de facto exceptions to this rule are the one or two men who work on Sundays feeding the fazenda cattle). This accounts for 3,645 man-days. Another three days were compulsory religious holidays, accounting for another 210 man-days. There remain 21,731 man-days. Of these 12,068 man-days are employed (as itemized in Table 11) in agricultural production in the workers' own plots; this accounts for much of the available labor time, leaving 9,663 days. Some of this remaining time is employed to fill labor obligations (*sujeição*) to the fazenda by maintaining fences, sitios, irrigation ditches, and the like. Sujeição labor employs 17 adult mules (from 15 households) two days each week for the whole year (1,768 man-days), leaving 7,895 man-days still to be accounted for.

Some of this remainder can be explained away as wage labor, but it is not easy to say just how much. The heads of 39 households could between them remember working 1,278 mandays for wages—either for the landlord of Boa Ventura (but not under sujeição), for other residents of Boa Ventura, or for neighboring landlords (see Table 12). But indications are that this figure is too small, probably because the workers interviewed could not remember every casual job they had done in the course of the year. In a separate inquiry (Table 13) I asked 32 workers exactly what they had done each day for ten days immediately preceding the interview; it developed that 12 per cent of the total working days available to these men had been spent in one form or another of wage labor. It is likely that the first 39 workers, together with the five married workers I did not interview, worked at least another 1,600 man-days for wages, in addition to the 1,278 reported.

So far, we have accounted for all but 5,017 of the 25,586 potential man-days of agricultural labor available on Boa Ventura. From the more intensive interviews (Table 13), I determined that the largest part of this remaining time is spent in sickbed, travel, or idleness. "Idleness" is a complex factor. A good part of it is outright unemployment, for many workers have little to do during the months of November, December, and January: they have harvested most of their crops, and they cannot plant until

TABLE 13
Allocation of Labor (Intensive Survey: 32 Workers for 10 Days)

Activity	Man-Days	Per Cent of Total
Labor in workers' own fields and fazenda obligations	182	57%
Wage labor	40	12
Ill	21	7
Travel,visiting	16	5
Resting (took a day off)	6	2
Household improvements	4	1
Could not remember	4	1
Sundays and holidays	47	15
Total	320	100%

TABLE 14
Market Value of Agricultural Production (44 Households)

Crop	Total Produced	Value	Per Cent of Total Cash Output
Maize	134,093 liters	Cr$20,113,950	31.5%
Beans	38,280 liters	17,226,000	26.9
Manioc flour	19,688 liters	2,953,200	4.6
Lima beans	9,480 liters	4,256,000	6.7
Potatoes	6,797 kilos	3,398,500	5.3
Rice	6,402 liters	2,880,900	4.5
Cotton	47,960 kilos	13,107,700	20.5
Total		Cr$63,947,250	100.0%

the start of the rainy season; the only available task is the clearing of new roçados. Then, there are some days spent in idleness by choice—a man simply decides to stay in his hammock for a day. A very small percentage of time (about 1 per cent of the total) is spent in making home improvements such as fencing a yard, repairing a roof, or adding on a new room, nearly always at the worker's own expense.

Labor output (productivity). The discussion in the following section will be limited to the production of workers in their own fields, so that it will be comparable to my estimates of agricultural labor input. I will deal with livestock and fruit production elsewhere.

In 1967, the Boa Ventura workers could sell their major crops in local markets for the following prices* (there are considerable price fluctuations during the year, but these prices refer to harvest time, when most crops are sold).

Maize ....Cr$150/liter   Rice.........Cr$450/liter
Beans...........450/liter   Cotton 273/kilo
Manioc flour .....150/liter   Potatoes.........500/kilo

By combining these figures with the estimated crop production of the 44 households in my economic census, I calculated the output of the workers' labor in terms of cash, or market, value. The results are given in Table 14.

* At the time of my study, a U.S. dollar exchanged for 2,700 cruzeiros.

We now know, at least approximately, the labor input of the Boa Ventura workers and the value of the marketable produce resulting from this labor; hence we can assign a value to a given unit of labor. This naturally relates to the topic of wages.* It should be remembered that this discussion refers to a period of good rainfall, when there were large crops and plenty of work. Had 1967 been a dry year, men would have worked very hard for very little, and there would have been a large pool of migrant labor to keep daily wages down.

The standard wage for a day of unskilled labor in the sertão is Cr$1,000 (37 cents) per day. In addition, as is the case in many parts of the world, the employer provides lunch and dinner. These meals are always cheap and simple: manioc flour, beans, crude sugar, perhaps some maize, and minute amounts of fat for seasoning; the total cost is about Cr$700 per worker. This rate of pay is by far the most common. However, wages do vary for a variety of reasons (although meals are always included, whatever the wage rate).

Wages vary with seasonal shifts in the local labor needs. As we have seen, there is no large migratory labor force in the immediate area of Boa Ventura, and almost all laborers are also residents of a plantation. They will often do extra work on neighboring fazendas; but, since everyone plants the same crops, the peak labor needs of both employer and employee coincide. Thus the labor supply is fixed, whereas the demand for labor varies, and wages rise and fall as the seasons change. During the cotton harvest, men are paid according to the quantity they pick, and a hard worker may earn as much as Cr$2,500 a day. Following this is a slack period, when many men simply cannot find wage labor at all; it is then that the going rate is Cr$1,000, for a man will not normally work for less under any conditions. But once the spring rains begin, the men have labor in their own fields to occupy them, and wages slowly begin to rise. By February and March they are up to Cr$1,500; later, in

---

* For a broader and more detailed treatment of this topic, see Johnson and Siegel 1969.

April, they decline to Cr$1,200. The wages of skilled laborers are less affected by the season, and are generally higher. A mason (*pedreiro*) or carpenter will often do piecework rather than accept a daily wage; but he can usually expect to earn at least Cr$3,000 per man-day.

The relation between a worker and his employer has a definite effect on his wage. The above wages pertain to work between two otherwise unrelated persons, and are consequently the highest. But workers who perform necessary labor for the fazenda as a part of the conditions under which they have access to land are paid only Cr$500 per day. Likewise, a skilled worker receives only Cr$2,000 per day (plus meals) for his work on fazenda projects. Furthermore, when a man works for a close friend or relative, he often works for Cr$100–200 less than the going rate.

The workers definitely feel that wage labor is not nearly as good as labor applied to one's own fields. It is not that such labor is despised (see Miller 1967: 174); rather, it pays far less. For this reason, the moradores are most willing to work when there is little left for them to do in their own fields. On occasion, a worker has pressing needs for cash that force him into wage labor; but this is always seen as a distinct hardship, and other possibilities, such as going into debt to the landlord, are explored before too much wage labor is undertaken. This reluctance to work for wages is based on firm economic grounds. The average cash value of one man-day in a worker's own fields is the total cash value of the produce of 1967 (Cr$63,947,250) divided by the total of man-days (12,068) required to produce it: that is, Cr$5,299. Even considering that about 25 per cent by value of the total produce is transferred to the landlord as his "share" of the crop, a man is left with an average return on labor in his own fields of over Cr$4,000—twice the value of wage labor.

A final question that may be raised at this time is: how does labor productivity relate to household consumption? More generally, how are the products of labor disposed of? The informa-

TABLE 15
Production and Consumption (44 Households)

Crop	Production (Estimated 1967 Harvest)	Consumption (1967)
Maize	134,093 liters	7,858 liters
Beans	38,280 liters	
Lima beans	9,480 liters	25,450 liters[a]
Manioc flour	19,688 liters	22,685 liters
Potatoes	6,797 kilos	6,797 kilos
Rice	6,402 liters	7,545 liters
Cotton	47,960 kilos	none
Total value	Cr$63,947,250	Cr$22,868,000

[a] Total consumption of beans and lima beans.

tion available on this so far is summarized in Tables 15 and 16. Table 15 shows the degree to which the fazenda is able to meet its basic food needs. Note that except for deficiencies in rice and manioc flour, which are made up by purchases in local shops, production more than equals consumption. Table 16 shows how the total production of crops is allocated.

From the preceding discussion of agriculture on Boa Ventura several definite conclusions have emerged:

(1) Rights to land are defined by the tenants in terms of the labor invested in the land, since the right to sell the land itself is in the hands of the landlord. In this regard, the moradores are rather like the members of a corporate landholding group who recognize individual rights of usufruct but reserve the ultimate right of alienation of the land to the entire group. In such a group, rights of usufruct are acquired by joining the group, whereas the moradores acquire their rights by agreeing to transfer a share of their produce to the landlord.

(2) A morador gains very little economic advantage by owning a small plot of land himself; and he may often work on a fazenda even if he does have his own farm. Nonetheless, he takes pride in his ownership of land. Like most workers, he is constantly balancing independence and pride with dependence and security.

(3) A considerable variety of crops and soils is available to

TABLE 16
Disposition of Products (44 Households)

Disposition	Value (Cr$1,000)	Per Cent of Total Value
Household consumption	22,178	34.7%
Maize and beans sold or fed to livestock	19,312	30.2
Maize, beans, and manioc given to landlord	9,349	14.6
Cotton given to landlord	6,554	10.25
Cotton sold to landlord[a]	6,554	10.25
Total value	63,947	100.0%

[a] The landlord is entitled to half of all the cotton grown on the fazenda and buys the rest at a reduced price.

the workers, and they exploit this variety in two ways: by selecting crops most suited to particular soils, and by spreading risks to reduce their insecurity in the face of a drastic and unpredictable climate.

(4) The workers' own classification of land types corresponds closely to their actual planting behavior, and probably derives from it. Anthropologists are constantly faced with the problem of how closely "native behavior" may correspond to "native categories," and it is unsafe to assume any one-to-one correspondence. In this case, however, a close correspondence has been demonstrated, and it is quite clear that decisions of where to plant which crops are made in terms of the land-types model.

(5) Planting behavior results from a complex set of decisions, and is to some extent idiosyncratic. Differences in behavior are largely caused by individual adaptation to special circumstances (e.g. available household labor, or household location). But to some degree, the idiosyncracies result from imperfect agreement on the real nature of things, which derives from the workers' lack of information about the probable outcomes of particular decisions.

(6) The agricultural technology of Boa Ventura is simple, but it is not necessarily the case that a transfer of methods from some "developed" agricultural system would benefit the work-

ers. Certainly, a morador would view such changes with well-founded suspicion until he had seen their material benefits for himself.

(7) The worker on Boa Ventura is idle relatively little of the time. He must produce for his family and for the landlord. He can always use cash; so, when his own fields are in satisfactory condition, he may seek a few days of wage labor. In a good year his hard work pays off. About one third of his total output goes for immediate food needs, and another fourth is paid in shares to the landlord. The remainder of the crop may be sold and used for purchasing clothing, tools, luxuries, or livestock.

Five

The Subsistence Economy: Specialization, Livestock, and Capital

The subsistence activities of the moradores of Boa Ventura are predominantly agricultural. Hunting and fishing do not occupy much of a worker's time, although the protein they yield may be significant to some households; nor do wild plants appear to be used, except in a few herbal remedies. Three nonagricultural aspects of the subsistence economy, however, are quite important. (1) Boa Ventura has a number of economic specialists who provide a diversity of goods and services. In this respect the fazenda differs from many other South American plantations, on which all workers are so alike that they must go outside the plantation for special products they themselves do not produce (cf. Miller 1967: 151). Special skills are difficult to acquire, however, and skilled workers are in a strong bargaining position. Off the fazenda, they receive higher wages than unskilled laborers can command; on the fazenda, they gain increased security as well as higher wages. (2) The moradores also raise livestock, which accounts for a considerable proportion of the total cash value of production on Boa Ventura. (3) Both the fazendeiro and the worker must invest a certain amount of capital in production. But in both cases the amounts are very small in comparison to those needed for technologically advanced plantations and some other peasant economies—to say nothing of European and North American farming.

ECONOMIC SPECIALIZATION

The following male specialists are found on Boa Ventura.

Masons (3). Masonry is skilled labor, involving few tools; however, a long apprenticeship is required to learn the proper use of the plumb line and level, the mixing of mortar, and so on.

Gardeners (3). These men must know such agricultural techniques as the application of fertilizer and insecticide, pruning, and irrigation for their work in the sitios of the owner.

Carpenters (2). Carpentry requires a heavy capital outlay in tools, and usually a long apprenticeship at very low pay. The carpenters do all the woodwork for the plantation, from shaping roofing beams, doors and windows (at fazenda expense) to building chairs, tables, chests, etc., for particular moradores.

Shopkeepers (2). Running a small bodega requires the ability to do mathematical calculations and accounting; it also requires capital and the ability to establish permanent trade relations with a reliable clientele.

Vaqueiro. The vaqueiro of Boa Ventura, under a special contract with the owner, is responsible for the care of the fazenda cattle. He applies various medications, trains horses for the difficult task of finding and chasing cattle in the dense, thorny brush, works leather for making saddles and protective garb, oversees the milking, and makes cheese.

Goatherd (*vaqueiro de bode*). The fazenda has a small flock of goats, but the goatherd's job is much less skilled and therefore less rewarding than the vaqueiro's.

Manager. To be discussed in Chapter Seven.

Blacksmith. This is the most nearly full-time profession found on Boa Ventura. The blacksmith not only repairs and sharpens the tools of the agricultural workers but also makes his own tools (files, hammers, etc.).

Brickmaker. Makes, dries, and fires the bricks used in house construction.

Carter (*freiteiro*). The carter owns a string of donkeys and a set of harness and leather baskets. As part of his sujeição obligation, he carts the owner's share of each harvest from the fields to the fazenda's central storage area. It is hardly skilled labor, but few workers can afford to keep even one donkey, let alone four or six and the equipment for hauling produce. When the freiteiro is not working for the owner, in fact, he usually loans his donkeys to friends, who repay him with gifts of produce.

Mechanic. The few machines the fazenda possesses (mostly pumps and generators) require maintenance and not infrequent repair. The mechanic does this as a favor to the landlord, and receives access to special lands in return.

This list includes enough professions to meet the most common needs of the moradores, and they do not have to run errands off the fazenda in most cases. There are shops off the fazenda that offer a wider range of goods at comparable or lower rates, however, and certain specialists (pharmacist, shoemaker, priest, schoolteacher, and folk healer) can only be found outside the plantation. On the fazenda itself, no specialty is carried out to the absolute exclusion of agricultural activity; even the masons, who have had years of training and are highly respected, also perform agricultural labor themselves. And because the specialists' earnings are unusually high, most of them can afford to hire extra labor at low wages.

There are seven female specialists on the plantation as well. Three own their own sewing machines and take in sewing for the local housewives, who buy their cloth in bulk rather than paying for ready-made clothes. These seamstresses have more work than they can handle, and usually do not occupy themselves much with household chores, which they delegate to their daughters, or even to hired girls. Three other women weave the straw bands that are used to make hats; this is simply a minor chore to earn extra money for the household, and is not particularly skilled. Finally, one woman does all the cooking for the landlord when he and his family visit; in return, she is allowed a special contract on her agricultural products (she is a widow who farms her own fields).

Special skills are acquired in no systematic or formalized way.

A few men, notably the vaqueiro, acquired their skills from their fathers, but this is not usual. Rather, it is felt that a man either has or does not have the personal makeup (*natureza*) to acquire a skill, and that he must follow his own natureza. When asked if they wanted their sons to be anything special, fathers universally answered, "It is up to him." To take a particular example, one of the carpenters now living on Boa Ventura had made up his mind as a young man to become a carpenter. He began, little by little, to acquire the tools he needed, and gradually trained himself by close and persistent observation of a skilled carpenter. Similarly, one of the masons began as an unskilled construction laborer, then worked for three years at a low wage while he picked up the skill from a friendly master builder.

It is not an easy matter to acquire skills; and there is no one to push, or even encourage, a young man who wishes to do so. Antonio, one of the younger married men on Boa Ventura, could see that his future, without skills, would be the dead end of unskilled agricultural labor. One day, while observing the construction of a building, I saw him ask the mason for permission to lay down a row of bricks himself. The mason assented without enthusiasm, and Antonio went ahead, asking questions as he proceeded. But the other laborers watched and jeered at him: "Do you think you are better than we are?" and, "You do not even know what you are doing!" He finally gave up, clearly affected by their gibes.

The skilled person occupies a different position within the plantation economy than he would outside. As I have noted, he is never exclusively a specialist, but always does some farming as well. In addition, when he does skilled labor for the plantation, he is paid less than he would receive for the same job elsewhere. All men who live on Boa Ventura are obliged to concede something as the price of their residence, and it is no different for the specialists. What the specialist does obtain is increased security. This is clear from the following history.

A young man, with five children, decided to move from a plantation to a town. He was a skilled carpenter, and he knew that wages in town were as much as twice the amount he was receiving on the plantation. But, when he moved to town, he suddenly found himself paying rent and buying all the family's food. He was financially secure as long as he could find work every day, but there were times when no work was available. Furthermore, his employers at times could not or would not pay him all they owed him. He realized that on a plantation, although the wages were lower, he would always have a roof over his head, and he would have crops to sustain him when no one needed his special skills. His experiment in town presented him with a loss (he had to sell all his animals and valuables in order to feed his family); and he left, after only six months, to settle on Boa Ventura.

LIVESTOCK

So far, I have said little about the place of livestock in the economy of Boa Ventura. In fact, the fazenda livestock are far more important to the owner than the agricultural products, including cotton; and raising animals is important to the individual moradores as well. By livestock, one means first of all cattle, for it is in them that the wealth of the plantation is most concentrated. At present there are 242 head of cattle on Boa Ventura, and the herd is constantly increasing. A small contribution to the plantation economy is also made by the herd of 57 goats (value Cr$712,500). But it is the cattle that contribute most to the dramatic difference in income between the landlord and the average worker.

The labor involved in cattle raising is quite small relative to the income derived from it. Nearly all the work is performed by the one vaqueiro, except during the last months of the dry season, when the herd must be fed daily, and during the annual roundup. The herd had a natural increase of 44 newborn calves during the year of this study. In addition, the owner bought 47 mature but lean cattle in order to fatten them for rapid sale at a profit. He sold all of the 47 and another 24 of the herd, and was left with a net increase of 20 in his herd.

TABLE 17
Worker-Owned Livestock (45 Households)

Stock	Number	Households Owning at Least One	Stock	Number	Households Owning at Least One
Pigs	199	40	Chickens	587	43
Goats	72	21	Turkeys	49	16
Sheep	48	6	Guinea hens	46	12
Donkeys	45	17	Ducks	21	10
Cattle	11	4	Peacocks	3	1
Horses	7	6			

Fattened cattle bring about Cr$300,000 per head in Fortaleza, so the owner earned Cr$7,200,000 from his 24 original cattle and netted another Cr$4,600,000 from those he bought, fattened, and sold. His total income was thus Cr$11,800,000 (US$4,370); and meanwhile his herd was increasing. At an average value of Cr$200,000 a head, the herd at the end of the year was worth about Cr$48,000,000 (US$17,777). Since the cattle feed for most of the year on either wild brush or the stubble of harvested fields, the owner's costs, over and above his initial investment in the plantation, were relatively slight.

By contrast with the considerable value of the owner's herds, the total value of worker-owned livestock is small (Cr$9,-355,000). However, there is some variety in the animals owned, as shown in Table 17. Pigs have the most economic importance, accounting for about a third of the total worker-owned animal wealth; and the average household owns 4.4 pigs (including piglets). A few workers raise pigs in order to sell them, usually as piglets; but most raise them until they are big enough to slaughter for home consumption (and for making gifts—a very important use of pork to be discussed in Chapter Six).

From the figures given by the workers themselves, it appears that raising pigs gives only a small profit at best, and may easily turn into a losing proposition. One worker estimated that five liters of corn must be consumed to produce each kilogram of pig. When a mature pig is sold, it brings Cr$650 to Cr$850 cruzeiros per kilogram, depending on how fat it is; yet five liters of corn usually cost about Cr$750 in the stores. Insofar as pigs fatten themselves by feeding on other sources, and insofar as corn invested in a pig tends to have a consistently high value (maize prices fluctuate more than pork prices, and often go well below Cr$150 per liter), it is to the worker's advantage to raise pigs. But insofar as pigs get sick and die uneaten, or must be sold thin and hence at a loss, it is not to his advantage to raise pigs.

There are about 13 chickens per household on Boa Ventura. They require relatively little care (and small amounts of maize), and are an important source of meat; but their primary value is in eggs. Unfortunately, diseases frequently exterminate most of the chickens on the fazenda. During this study, one such epidemic hit, and chickens were simply unavailable for two months.

The raising of horses, cattle, and donkeys is a highly preferred activity, but there is a great barrier to it. Whereas pigs and chickens do not consume fazenda resources that could be applied elsewhere, the larger grazing animals compete directly with the landlord's own cattle for the limited pasture available.* The workers who own these animals must keep them penned up and provide them with food, and this is always costly in either labor or cash. To raise large animals at all, one must get permission from the owner, and it may not be granted if he thinks the animals will be grazing on his pasture.

Finally, dogs and cats are found in every household, but have no economic importance.

There are important differences in the animal husbandry practices of the various moradores of Boa Ventura, as there are in every other economic activity on the fazenda. Some families raise no livestock whatsoever; the vaqueiro, by contrast, owns stock worth more than a million cruzeiros. More than half of the workers own less than Cr$150,000 worth of livestock, but only ten own more than Cr$200,000. This does not mean that

---

* Miller (1967: 172) reports a like situation on a Peruvian plantation. Before the owner started raising cattle, the Indian workers each had a small-to medium-size herd of cattle; now they have none.

the households with very few animals are necessarily the poorest; they may prefer to store their wealth in bicycles, sewing machines, or radios. On the whole, however, the greater the value of a worker's livestock, the better off he will be found to be.

CAPITAL

Fazenda capital. As must be abundantly clear by now, only small amounts of capital are needed to keep the fazenda going at its present level. Boa Ventura, like most other fazendas throughout Ceará, could not exist without an enormous initial investment in reservoirs, which make water available all year round and for several drought years in succession if necessary. But most reservoirs were built many years ago, and the last reservoir work was done on Boa Ventura in the 1940's when the height of one of the dams was increased. Some additional irrigation apparatus has been installed in the sitios (where few workers plant crops). Basically, this is a series of irrigation canals, constructed of locally made materials according to a simple design; two gasoline-powered pumps have been installed to move water during the dry season.

Over the years, several buildings have also been built for fazenda purposes: a manioc flour mill, the cattle stable, a sugar mill (not working, but restorable), and various other structures connected with production. Few of these require any expenditure beyond small amounts for maintenance. Resident housing, however, is another matter. Close to 50 houses now stand on fazenda property, and, if 1967 was any indication, about five must be constructed every year to replace old ones. By custom, the owner provides housing, so this expense is his.

Most of the improvements found on Boa Ventura were made by previous owners. When the present owner bought the fazenda in 1962, he moved one of the gasoline pumps that had been used for irrigation to another fazenda, thus reducing the total area that could be irrigated at any one time. The largest investment that he himself made was the purchase of a large flatbed truck for transporting cattle and produce.

TABLE 18
Workers' Investment in Tools

Tool	Number on Fazenda	Cost
Hoe	151	Cr$453,000
Large billhook	89	267,000
Small billhook	80	240,000
Ax	42	189,000
Large knife	11	27,500
Total cost		Cr$1,176,500
Average investment per household		Cr$27,400

Workers' capital. As I mentioned in Chapter Four, the mora-dor can perform all his normal tasks with a very small inventory of tools. The total possession and value of the basic work tools on Boa Ventura is given in Table 18. Other tools are sometimes found—picks, shovels, rifles, etc.—but these do not raise the average household value above Cr$32,000 (less than US$12).*

For the majority of the workers on Boa Ventura, the largest permanent investment is in tools. However most households also invest in one or more pigs, which offer a way to store capital in a relatively liquid form. In Ceará, as elsewhere in Brazil, in-flation drives prices up at the rate of about 40 per cent per year, and cash savings lose value rapidly. The pig is thus a savings bank (or living "piggy bank," as Ralph Beals calls pigs used similarly in Oaxaca, Mexico).

A few moradores store their capital in manufactured items of some value, such as radios. Here again, there are different pri-orities, and most workers make distinctions between items ac-cording to the value of services produced. For example, a bi-cycle, which costs about as much as a radio, is thought to be

---

* This is further evidence that the technological simplicity of Boa Ven-tura is only incidentally related to a shortage of capital, since workers have a considerable amount beyond consumption available for reinvestment. That only enough reinvestment takes place to maintain the small inventory of tools merely indicates that other expenditures—for clothing, pig raising, or the purchase of luxuries—are seen as more desirable or more feasible than improving agricultural techniques through investment (see pp. 72–76).

more useful. One informant had bought a radio because he had been offered a good bargain, but he was looking for the opportunity to exchange it for an item productive of more tangible services. Only three other men owned radios; for them, a radio was a prestige item, and they had only purchased one after many more practical valuables had been obtained. The same reasoning is applied to livestock other than pigs: thus a burro is considered a better investment than a horse because it can be both ridden and used as a draft animal; and unlike a horse, it needs no special feed.

Three moradores are oriented toward the use of cash in hiring labor. The amounts they employ are very large by their own standards—from Cr$300,000 to over Cr$1,100,000. Two of these three "entrepreneurs" borrow the bulk of their cash from the landlord; they must pay him back (though without interest), and must give him shares of their harvest like any other moradores. Unlike most moradores, they are taking a great risk, since a fairly poor year, let alone a drought, will deny them the wherewithal to pay off the debt.

Although the moradores who actually invest capital (beyond the minimum replacement costs for tools and other necessities) are few, those who put aside some form of savings are numerous. The pig is one form. Another is the storage of staple crops far in excess of a household's needs before the next harvest. A hoard of this kind is considered very desirable, but it is difficult to maintain. Moradores usually find themselves with an excess of crops after the harvest, but are forced to sell them during one emergency or another, until they run out. About three-quarters of the moradores had run out of beans, corn, and manioc flour before the 1967 harvest had begun. The storage of a crop surplus is considered the best security against a drought year. When workers see that the new harvest will be bountiful, they sell their old surplus; this results in the curious phenomenon that staple food prices go down *before* the harvest of new crops has begun. Obviously, the workers are capable of planning and deferring consumption; deferring consumption in order to make risky investments, however, runs entirely against the interests of most of them.

It should be apparent from the discussion in this chapter that the investment in skills or productive capital needed to keep Boa Ventura going is extremely low, at least by the standards of an industrial capitalist economy. Within the context of the fazenda, however, the differences between moradores in these matters appear rather large. Skilled workers have somewhat greater prestige and more security than unskilled workers; wealth differences are reflected to a considerable extent in the number and kinds of animals owned; and basic differences in outlook seem to separate the "entrepreneurial" users of capital from the average morador.

It should perhaps be pointed out here, in conclusion, that despite the existence of considerable economic differences between moradores, the workers themselves do not attach much significance to these differences; that is to say, they do not believe that important prestige differences separate them from one another, nor do they employ toward one another the deferential modes of stance and speech that they employ toward persons of high status, such as the landlord. Whatever differences of degree exist between moradores are far overshadowed by the differences of kind that separate them from the higher classes.

Economic differences between moradores are in matters such as sleeping in beds rather than hammocks, owning one horse rather than none, wearing pants with few patches rather than many, having chairs rather than a log bench for guests, and so on. Differences between landlords and moradores are far more extreme: owning a large brick and plaster house of 10–15 rooms rather than living in a rough mud or brick home of 2–3 rooms owned by the fazenda; owning a truck; having a high school education; wearing expensive store-bought or tailor-made clothes rather than locally sewn clothes of cheap cloth; and so on. In dealings between one morador and another, one finds a mutual sense of equality; in dealings between a morador and the landlord, one finds a strong mutual sense of inequality.

Six

Economic Relations: Horizontal

Despite the rather wide differences between moradores, they consider themselves as equals when compared with the landlord, the manager, and other members of the local upper or middle classes. Naturally enough, the economic relations between moradores are quite different from those between moradores and higher classes. The "horizontal" relations involving people who are considered equals will be discussed here, leaving "vertical" economic relations to the next chapter.

The kind of horizontal exchange relation to be discussed here is found between both kin and non-kin; however, close exchange ties occur with greater frequency between kin than between non-kin. The real, critical difference is between the economic relationships within a household and those between separate households (cf. Sahlins 1965: 141). The members of a household are united by common interests, under the authority of the dono de casa. They make demands on each other according to definite rights and duties that they recognize toward one another; and in most respects, they act as a unit. By contrast, the members of separate households, even if they are close relatives, do not recognize inherent rights and duties toward one another. There are ways of creating these rights and duties; but they depend on the volition of both parties, and they may be dissolved by either party quite easily. The kinship relationship, where it is also an economic relationship, is the model for all close economic relations between households. On Boa Ventura, the formation of economic ties between households results from a drive for security and persists through the mechanism of mutual aid. Since close kinsmen usually take the responsibilities of mutual aid more seriously than non-kinsmen, there is a tendency for close horizontal exchange relations to involve kinsmen.

Cyril Belshaw has remarked that "all enduring social relations involve transactions, which have an exchange aspect" (Belshaw 1965: 4; cf. Sahlins 1965: 139). Horizontal exchanges on Boa Ventura differ from vertical exchanges in that both partners give and receive similar items. However, these horizontal exchanges are not simple one-for-one trades, but correspond in form to what Sahlins has called generalized reciprocity. "The material side of the transaction is repressed by the social: reckoning of debts outstanding cannot be overt and is typically left out of account. This is not to say that handing over things in such form, even to 'loved ones,' generates no counter obligation. But the counter is not stipulated by time, quantity, or quality: the expectation of reciprocity is indefinite." (Sahlins 1965: 147.)

Although the expectation of reciprocity here is not explicit, that expectation definitely exists. In fact, the moradores' strong expectation of reciprocity at some future date approaches Sahlins's concept of "balanced reciprocity." But, as we shall see in Chapter Seven, the relation between a worker and his landlord is even closer to a state of balanced reciprocity in the minds of the workers.

HOUSEHOLD ECONOMICS

Nominally, the father of a family is the dono de casa, and all economic decisions rest ultimately with him. In practice, because his actual power is not absolute and because he himself may not wish to, the father does not make all the economic decisions. His most common decisions are in the allocation of household labor and in making purchases. The father is the only family member with a clear-cut obligation to work in the fields. He supervises his family's labor there, and he himself puts in the most work. Besides planning and executing the year's agri-

cultural tasks, he must acquire or save seed, maintain and replace tools, borrow or earn cash when needed, make nearly all
purchases in the stores, and represent his household with the
landlord or the manager.

A wife has the definite obligation to obey her husband in
economic matters. There are several spheres in which women
generally do not interfere. Many women have practically no
knowledge of money, and therefore do very little business in
the local bodegas. Many women, too, cannot help with the field
work (although husbands often need their help in the lighter
tasks of harvest time) because they have young children who
must be cared for. In any event, women are expected not to
interfere in the decision of what to plant. In 1967, one man's
wife urged him to plant rice in his roçado, arguing that the winter might be wet. In keeping with his role as agricultural decision maker, the husband decided not to plant rice because it
would require extra labor and involved a high risk of failure. As
it turned out, it was a very wet year, and those who did plant rice
were well rewarded. The worker told this story in a mildly
humorous manner, pointing out that his wife had guessed right
—but with no admission that he ought to have followed her
advice, in 1967 or in any other year.

Despite these limitations, a wife does have some spheres in
which she is relatively free to make decisions. She decides what
to cook, and when to sweep the house, and perform other domes-
tic tasks; she also oversees the livestock, poultry, and household garden. If her husband does not require labor of the older
children in the fields, she may use them for her own purposes.
And when a person comes to the house to buy or sell some small
item, she may make the transaction without first consulting her
husband. More important than any of these responsibilities,
however, is the wife's practice of maintaining friendly relations
with neighbors by exchanging small amounts of food. Many
husbands insist that they have the last say about these exchanges, particularly exchanges of meat; but I noticed that
women would generally take it upon themselves to give gifts

and would advise their husbands only afterwards, if at all. One man said that his wife often made gifts secretly, without telling him, but that he did not mind this because he had confidence in her ability to do it well. This practice fits in with the wife's general household function: through such exchanges, she can acquire foods she does not already have by giving away something else she has in abundance; hence she is able to give her family a more varied and nourishing diet.

The children, like their mother, are first of all obliged to obey their father. After the age of about ten, they are expected to make their labor freely available to the household. Girls of that age are already entrusted with the care of younger children, and with such household chores as carrying water and sweeping. Sons often (but not every day) accompany their father to the fields and do simple weeding, planting, or harvesting chores. The older the boy, the heavier the demands made on him; and when he is about 15, he is expected to be fully as capable as an adult. But neither sons nor daughters actually behave as adults until they are married. They frequently shirk housework, wait to be told what to do, and at times do not work for days. This will vary depending on the family; many fathers demand strict obedience from their children, and some claim that their sons work for the household every working day.

Grown but still unmarried sons may act as substitutes for the father, under his authority: for example, a father owing a man-day of labor to another household may pay it off as easily with his son's labor as with his own. Whatever money a son earns he turns over to his father, and the crops he plants also belong to his father. In turn, the father gives his son clothing, tools, and food, and occasionally releases extra cash for luxury items like watches. A son who is about to be married may even be permitted to keep a part of his produce for himself.

Several households on Boa Ventura are extended to include a single person, whether a kinsman or not (see pp. 26–27). Those who are not kin are usually men, and they are subject to the authority of the head of the house, just as a son is. They are told what to do by the dono, whom they call a "patrão," and do not work for anyone else without his permission. Being subordinated in this way is unpleasant to these men, as it is to most moradores; and they generally stay only for short periods of great personal need.

When the extended family in a household is composed of kin, one of three economic relations is possible. First, newly married sons or daughters of the dono may be living with him until they can find houses of their own. In this case, the younger couple plant their own crops and frequently are not subjected to the dono; they may even have a separate kitchen where the wife cooks for the two of them. Second, there may be a dependent kinsman or kinswoman who has nowhere else to go, and who contributes what he or she can to the household, as ordered by the dono. Finally, there may be a sort of "ambulant youth" (male or female) who moves from one of his relatives' households to another, never staying in one household very long (always less than a year). As long as the youth stays in the household, he is subject to the dono. He may leave because he wants to leave and has found someone else to take him in, or because his relative no longer finds him useful. Some of the moves by a youth are clearly ways of filling labor needs, but others appear to be made for noneconomic reasons (e.g. to avoid conflict with a neighbor).

NON-HOUSEHOLD KINSMEN

All but 18 per cent of the households on the fazenda have traceable kin ties with at least one other household on the fazenda; and nearly all these ties are primary (parent-child or sibling), although some first cousins, uncles, aunts, nephews, etc. are found. A few other kin ties are recognized by the moradores, but the exact relationship cannot be traced. In general, households have primary kin ties with only one or two other households, but two larger "primary groups" exist: a widow and her six married children; and a couple and their three married children. At least 22 households have primary kin off the fazenda but within walking distance; they maintain some form of exchange with these relatives, which is generally how I found out about them. It is significant that many of the families involved in all this are not native to the region of Boa Ventura. Quite often, one family had come there first, and was followed by a near kinsman with his family.

Although the presence of a close kinsman is a definite contribution to a morador's security, specific obligations between kinsmen are not recognized. The most explicit obligation one finds is the moral obligation of a son to help his elderly parents, and even this strong expectation is not always realized. The factor most responsible for this weakness of kin ties is simple geographical distance: there is so much migration among the people of the sertão that it is common for a son to be too far from his parents to carry on effective relations with them, or even, at times, to know if they are alive. I found convincing examples of this lack of specific obligations even between nuclear kin. One worker reported that when he had lived on a piece of land owned by his brother, he had sujeição labor obligations just like any other morador; and that when he fell ill, his brother neglected to help him. Likewise, I saw two brothers sell various items to each other at going rates—that is, they did not think it improper to treat each other as strangers in a business transaction.

Nevertheless, there are definite ideals that do emphasize kin cooperation, even though they have no coercive force. The average morador prefers to live near a kinsman if possible; and all informants agreed that kinsmen will usually disagree less and find less to fight about, and hence that kinsmen make better neighbors. For this reason, there is a strong tendency for primary kin to live near one another. The "primary group" of a father and three married children mentioned above is an exception, for two of the children live some distance away; however, at the time I left the fazenda these two had definite plans to move next to their father the following summer. The only real exception to the rule was a man whose father lived on the other side of the fazenda. So it can be taken as a rule that a worker will live close to his nuclear kin if they are on the fazenda at all.

Thus one finds localized groups of primary kin scattered over the fazenda. But within these groups, relations may not be equal among all members. Among even nuclear kin there is a certain selectivity, based on the ubiquitous principle of proximity and the ambiguous principle of "friendliness." For example, the widow and her married children all live within less than ten minutes walking distance of each other; yet two of the children have by far the closest relations with her. Another child is subjectively very close to these three, although his household is the most distant, because the only path to any other houses brings him very close to the widow's house. On the other hand, two other children of the widow have close neighbors who are not kin, have less reason to pass near the widow's, and are the ones with whom the widow and the first three children exchange least.

Visiting appears to occur most often between the women, who have a good deal of leisure during the day (provided they have no special skills). Visiting takes place for the most part between kin. A woman who does not have a local kin group may still visit her neighbors, but usually stays at home more often. Men tend to do their visiting at night. In my observation, they visit over a wider area than their wives: that is, they visit more non-kin.

What differences in exchange patterns are found between kin, and non-kin? In fact, there seem to be no categorical differences, for the same types of goods and favors are exchanged. A member of one localized kin group described kin exchange in the following terms: "Our family is united; when one of us gets sick, all of the others are right there to help. We like to live next to each other, so that there are no fights." As far as exchange is concerned, a kinsman is the "good neighbor" par excellence. In one case that I observed a worker who was too sick to leave his hammock for over 50 days received help from every part of the fazenda, including the wealthy vaqueiro, the manager, and many far poorer workers; all of his relatives, except one of his brothers, also helped him out. For a period of many weeks, he lived exclusively on the charity of these people, strangers as well as kin. Informants will state that a kinsman can be relied on, whereas one takes one's chances with a non-kinsman as a neighbor. But the practical difference between kin and non-kin is more one of degree than one of kind.

ECONOMIC RELATIONS BETWEEN "EQUALS"

We are dealing here with exchange only between moradores, that is, between persons of roughly equal status: in particular, we are dealing with exchanges that seem to imply a special kind of relationship between the two partners to the exchange. It is common for one morador to sell another some item for cash, but this is simply an isolated transaction that implies no special relationship between the two men. On the other hand, an exchange of goods or services for some other goods or services, without the use of cash, often implies a more lasting relationship.

During the early days of my fieldwork, Fazenda Boa Ventura seemed to me an agglomeration of isolated households. But as the study progressed, I began to notice the contacts between households. Small exchanges, taking perhaps only a few minutes and performed without ostentation, were constantly occurring. Women came to visit and spun cotton for their hostess as they gossiped. Liters of beans, folk remedies, pieces of pork fat, cigarettes, squashes, the loan of a milk goat for an infant, piglets, and a host of other items appeared endlessly as "gifts." It is clear that a great many goods and services were being exchanged.

Upon inspection, too, it was clear that these goods are not exchanged in a haphazard fashion. First, there is a category of goods that are never exchanged as gifts, such as bicycles, radios, watches, and sewing machines. Goods of this kind are expensive; and when they are exchanged, it is always a business transaction, usually between relative strangers, and is expressed in exact money values. The common necessities of life constitute another major category of goods. They are given as gifts, but never as unsolicited gifts: one does not send beans to a friend who has liters of beans himself. Even a solicited gift does not always involve an open request, although open requests are common and no stigma attaches to them. Frequently, a man indirectly informs a close exchange partner that he is out of something that his partner possesses in quantity, and the partner responds with a gift. For example, when I questioned one of my informants about gift exchange, and especially exchange between friends, he later brought me a gift. In his terms, I had broadly hinted that he owed me a gift if we were actually friends. A third category of gifts includes what can best be called small luxuries: these are mainly foods such as meat, tapioca, fruits, and vegetables. They are sent as unsolicited gifts, for they are rare enough that one can assume his exchange partners are probably without them.

Two types of exchange items, labor and meat, correspond, respectively, to the second and third types of gift. A closer examination of these two items will point up some of the differences between them, and will illustrate the phenomenon of gift exchange as a whole among the resident workers of Boa Ventura. But before we pass on to these specific examples one final point must be made. With the exception of labor exchange, the exchanges that imply personal ties between households are handled by women. In this important area, the men, nominally the ultimate authorities of household economic policy, are in fact left out. Since these exchanges are a kind of insurance against want, as well as a promise of small amounts of luxury foods throughout the year, they put an important sphere of economic influence in the hands of women.

Labor exchange. There are two common ways for a morador to acquire extra labor when it is needed in his fields: hiring and exchange. Of the two, hiring is much less common. Only a fifth of the heads of household on Boa Ventura hired labor during the year of my study, but three-fifths exchanged at least one day of labor. The deciding factor is apparently that hiring labor requires cash on hand, whereas exchanging labor does not.

Labor exchanges usually occur between persons who have established an exchange relationship by other means (e.g. gifts of meat). Like most gifts, they are related to the needs of one or both participants. At certain points in the seasonal cycle, it may be very advantageous to rapidly complete a piece of work that a man could have done more slowly by himself. Suppose that a worker is still clearing his roçado while all the men who are clearing around him are finished and ready to burn their fields. They will burn even if he is not ready, in which case his plot will burn unevenly; and if he waits much longer, the rains are likely to ruin his chances for burning that year at all. In such a situation, the worker will try to find several others who are willing to put in a day's labor apiece and speed up the job; but there are no automatic and ascribed relations to guarantee that he will find this help. By going out and soliciting help, he sets up a reciprocal obligation: at some future time, he is expected to return the day of labor to each of those who have helped him.

This reciprocity holds for most "gifts," even when they are only small amounts of staple crops (although then accounting may be more casual). What makes labor exchange a form of "gift" rather than a simple economic transaction is that the one who gives the labor often does not benefit at all by the bargain, but does it as a favor to a man who is otherwise tied to him through a broader range of exchanges. In interviewing the workers who did and did not engage in labor exchanges, I found a general agreement that the exchange of labor is not of equal benefit to both sides; this is particularly true when the donor is a man who can himself afford to hire labor.

Meat exchange. The exchanges I recorded in this category are highly specific. The more informal exchanges of small staple foods and common items were so difficult to observe systematically that they do not appear in my data; but they were many times more common than exchanges of meat. Meat exchange patterns, by contrast, could be established quite accurately,

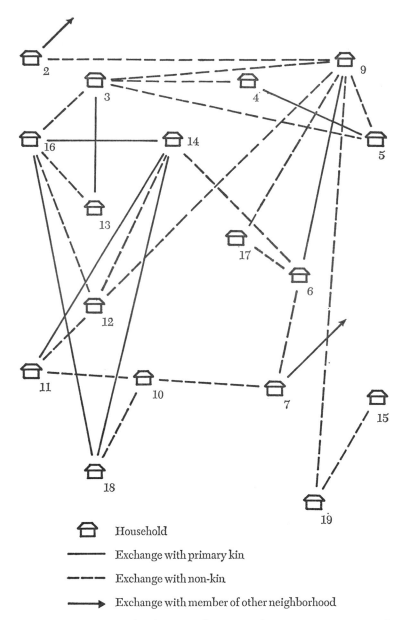

Household

——— Exchange with primary kin

----- Exchange with non-kin

——➤ Exchange with member of other neighborhood

Figure 6. An example of meat exchange: a schematic representation of one neighborhood, with spatial relations approximately preserved. (Houses 1 and 8 in my numbering of this neighborhood were unoccupied and are not shown.)

since the exchange of meat is an important event and is easily remembered by informants.

The usual chain of events begins when a household slaughters a pig or goat. That same day, it sends gifts of meat to several other households (usually from three to five). There is no immediate gift return or attempt at repayment with some other commodity; in fact, it is characteristic of these gifts that they are returned in kind. In the case of necessities (other than labor), one may return a gift of beans with a gift of potatoes at a later date and expect the exchange relationship to be maintained. But with meat (and with tapioca, another luxury), one waits, expecting a return of the same item in approximately the same quantity. The wait may be as long as several months without violating etiquette.

I recorded all meat exchanges between households made during the year of my study, and found a definite clustering of exchange relationships. Figure 6 shows the largest neighborhood on Boa Ventura, which contains 18 houses and lies south of the larger reservoir: here, households 3, 16, 13, 14, 12, 11, 10, and 18 make up one exchange cluster; households 3, 4, 5, 9, 17, and 6 form a smaller group. These clusters are spatially separate from one another; but since kinsmen tend to live near one another, it is not possible to say whether kinship or proximity does most to determine the occurrence of an exchange relationship.

A look at the facts of exchange behavior shows that both kinsmen and non-kinsmen form meat-exchange relationships, and that three general rules are observed. (1) All primary kin living on the fazenda exchange meat. Only two exceptions to this practice occur. Those few cases where meat exchange extends over a considerable distance on the fazenda always occur between primary kin; in Figure 11, the primary bonds leading away from households 2 and 7 are the only exchanges not contained within the neighborhood. (2) All neighbors exchange meat. There are more exceptions to this than to the first rule. Nevertheless, nearly all immediate neighbors exchange meat with one another, and the frequency of exchange ties diminishes as distance increases. It is hard to define just what a "neighbor" is in this context. The Portuguese word *vizinho* refers properly to proximal residence; but the Boa Ventura workers use the verb *vizinhar* (literally, "to neighbor") to refer to the exchange activities. When a worker sends meat to someone a few houses away (which would almost invariably be several hundred yards, considering the great distance between houses), he may receive nothing in return, in which case he will say that the other refused "to be a good vizinho." (3) Unrelated workers who live some distance apart do not form meat-exchange relationships, nor do they form any other sort of reciprocal exchange relationship.

General characteristics of exchange. The exchanges we have been looking at are invariably reciprocal. Informants are very clear about this. Said one: "I give meat to those who give meat to me. It is a loan; I send them so much meat, so that they will send me the same amount later." When a new neighbor moves in, many workers will attempt to set up an exchange relationship by sending him some item as a gift. If they get no response, they may try again; but if nothing happens after a few more tries, they will consider the new arrival a miser and will stop sending gifts of any kind.

Some valuable ends are achieved by the reciprocal exchanges on the fazenda. First, there is a constant redistribution of goods, as when a man who is out of beans receives a gift of them and reciprocates with potatoes. Second, luxury foods are more evenly distributed throughout the year: the man who slaughters an animal and gives away meat is setting himself up to receive return gifts of meat throughout the year, at times when he himself may have neither an animal to kill nor money to purchase meat. In the case of meat especially, the dietary advantage of fresh meat over salted or dried meat is being maintained. Third, the moradores recognize that by setting up exchanges two families have established a close relationship with one another—a relationship that they enjoy, and one that assures each family of having friendly and cooperative neighbors. If friendliness is lacking, both neighbors will feel uncomfortable, and disputes are practically inevitable.

Reciprocal exchange, as we have seen, is intimately related to kinship. So far, I have emphasized the lack of any "coercive" obligations between kinsmen. Now I wish to emphasize that in actual practice kinsmen are still important sources of security to each other, for they always establish continuing exchange relations. As one informant put it, gift exchange between kin is an "expression of unity." Physical proximity is as important as kinship in determining exchange partners, and we have seen that the two factors usually coincide.

One exception to the determination of exchange partners by kinship and proximity is very revealing. The vaqueiro, who holds by far the most wealth in animals and household possessions of all the moradores, receives gifts from a wide range of people, most of them non-relatives who live on other fazendas. During my study he received, among other things a shirt, two sheep, a saddle, a goat, and three pigs—all substantial gifts, and mostly from other relatively well-to-do persons, such as the managers, owners, or vaqueiros of nearby plantations. Although he had not given as many gifts in return, he had also given gifts to many other non-relatives off the fazenda.

The vaqueiro is very much like the manager of Boa Ventura, and like the other notables of the region, in that he maintains close relationships over a wide area through the mechanism of gift exchange. The owner of Boa Ventura, who is worth perhaps 75 times as much as the vaqueiro, does not enter into this circle of exchanges to any great extent, having his own circle in the urban center of Fortaleza. Reciprocal gift exchange carries a strong implication of equality between the two partners to it. The poorer moradores exchange with each other, but not with rich people. The wealthier peasants exchange with local shopkeepers, plantation managers, and small landowners. The great landlords, far above this middle sector in wealth, have their own elite circles for carrying on reciprocal exchanges.

Obligations to the very poor. So far, I have limited this discussion of exchange among social equals to reciprocal forms. Charity, however, is not reciprocal, and it is quite important. All workers recognize a generalized obligation to help someone who is too ill or weak to help himself. The gifts given in this context are, of course, necessities; and they are given unconditionally, as outright gifts. The moradores see this obligation as a kind of insurance for themselves: "If someone comes to my house asking for aid, I never refuse him; for I may someday need aid myself, and it will be given more freely if I have been generous." Blind beggars follow regular routes from fazenda to fazenda, calling from time to time at the same households and receiving small gifts such as potatoes or manioc flour. And I have already mentioned a worker who fell sick and was unable to leave his hammock for 50 days. His relatives nearby were few; but the manager of the plantation permitted him to continue living in the same house without making up his work obligations, and about a dozen other moradores sent constant supplies of beans and manioc to his home throughout his illness. The worker felt great gratitude for all this, but no obligation to "pay back" what he had received.

Debt and credit among social equals. Most debts among the moradores are contracted with persons of higher status, and these will be discussed in the next chapter. But workers occasionally borrow from each other as well. In all such cases, the amounts are extremely small, and the duration of the debt is very short: loans of less than a day's wages (i.e. about US $0.25), paid back within a few days or weeks, are the most common. The workers on Boa Ventura are not usurers, and frequently they charge no interest at all. When they do charge interest, it is typically about 5 per cent a month; considering the inflation of more than 3 per cent a month, this figure scarcely represents a profit for the man who receives it. In fact, none of the fazenda residents, worker or not, earns a significant proportion of his income by lending cash.

Willingness to make a small loan apparently does not imply that a strong reciprocal exchange bond already exists. I recorded

30 loans between moradores, but only 13 were between people who exchanged meat. Moreover, only one of these loans involved interest, indicating that the loans were made out of courtesy, not for profit. Obviously, the existence of strong exchange partners does not preclude exchanges and mutual aid between moradores on a more casual basis.

Barter (troca). Even though cash is scarce on Boa Ventura, barter is relatively uncommon. Most of the workers feel that they have no knack for barter, and give examples of bad trocas they have made in the past, remarking that "someone is always the loser in a troca." Nevertheless, barter does occur. Some moradores, in fact, pride themselves on how clever they are at troca, and they, too, give examples. But the critical fact is that troca is invariably expressed in cash terms: both parties know the approximate cash value of the items involved in a trade; and if these items are not strictly comparable, the trader of the less valuable item usually makes up the difference in cash. Since moradores seldom have large amounts of cash on hand, the acquisition of very valuable items is normally accomplished by troca. Money may not change hands at all in these trades, but there is no doubt that the trade items are substitutes for money (i.e. they have "money functions") rather than barter items in the strict sense of barter as a pre-money economic activity (cf. Belshaw 1965: 9-10).

Seven

Economic Relations: Vertical

In the last chapter the backlander's search for security was seen in one aspect: the formation of enduring bonds of exchange between the more or less equal members of the morador class. We will now see how the morador's search for security leads to enduring relations between members of different economic classes, i.e., to vertical economic relations. Vertical ties, like horizontal ones, are only formed when both parties are willing, and they are as easily dissolved. In vertical relationships, however, the exchanges are between unequals, with the result that the goods and services offered by each of the two partners to the exchange are different (cf. Foster 1963: 1281).

The security involved in horizontal relations arises from the fact that goods and services normally produced by the worker, such as labor or beans, are made available to him when he temporarily lacks them. Vertical ties, on the other hand, satisfy essential needs that the average worker can never satisfy on his own: the loan of relatively large amounts of cash or goods; the use of land, particularly the most productive land; medical aid; and general protection in unpredictable or catastrophic circumstances.

As is the case with horizontal exchange, no sharp line separates casual vertical transactions from personalized, enduring relationships; yet the distinction between the two is clear enough. Thus a worker may start buying in several shops when he first arrives on a fazenda; in each, his credit possibilities are nonexistent or extremely limited, and he shifts his business between them. In time, he begins to patronize one shopping place exclusively and his credit possibilities there are then limited only by the wealth of the shopkeeper. Similarly, the distinction between the present owner of Boa Ventura and the previous owner, which I will detail later, rests principally on the willingness of each to shift transactions with his workers from casual business affairs to a personalized economic and social relationship.

The worker himself conceives of his relationship with the landlord as one of exchange: he provides the labor and agricultural know-how, and the landlord provides arable land and some degree of long-term security. This picture closely resembles the situation that Sahlins (1965) describes as "balanced reciprocity." However, neither Sahlins's formulation nor the workers' own conceptualization necessarily implies that the exchange in this situation is "equal." There is a strong tendency among social scientists to consider "exchanges" as by definition equal; if an exchange is materially unequal, then some force has been applied, or some intangible, e.g. prestige, has also been transferred to "balance" the exchange (Homans 1958; Burling 1962; Le Clair 1962: 1182, 1184). Sahlins (1965: 146), however, proposes quite correctly that we study the material goods or services exchanged; and that rather than postulating some intangible, nonmeasurable quantity like prestige to balance the exchange, we should accept a condition of "one-sidedness" if such actually seems to be present. I have not been able to measure the exact value of the resources each party brings to the exchanges to be discussed here, but the reader should keep in mind the possibility that many of these exchanges are not at all equal.

One important vertical relationship falls outside the scope of normal subsistence transactions on the fazenda. Many of the workers on Boa Ventura perform wage labor for neighboring landowners. This is a significant area of economic life for the residents of Boa Ventura, since much cash income is derived from it. The relationship itself, however, is by and large an impermanent one, in which a day's labor is rewarded immediately with wages and meals. At the end of a day, or even at the end of a half-day, either party may terminate the arrangement; and the employer recognizes no other obligation than to pay the worker for his completed labor. Occasionally, a man who works frequently for the same employer may receive loans of money or food on the promise of future labor. But no sense of indebtedness, loyalty, or alliance exists in the relationship.

SHOPKEEPERS

In the bodegas that are interspersed about the area of Boa Ventura, usually beyond its borders, loyalties are the rule: enduring ties are formed between a morador and a particular shopkeeper, to the benefit of both. These ties are somewhere between horizontal and vertical, as so far described. Characteristically, a new morador moves into the area and begins shopping at various bodegas. When the need arises, he may ask one shopkeeper for a few days' credit; this is usually given, since short-term loans for small amounts are very common. If credit is not given, perhaps because the shopkeeper does not want any more debtors at that time, the worker will ask other shopkeepers until one agrees. And as long as he is in debt to a given store owner he will not shop at any other bodega. This is a fundamental rule of behavior, with strong moral overtones; as one informant put it, "I don't feel right shopping in one store and owing in another."

Once the worker has paid his debt in a bodega, his obligation to shop there exclusively is gone, and he may begin to shop at other stores again. But the usual tendency is to continue shopping where a successful credit relationship has been established. Even in this highly mobile population, I found one morador who had shopped at the same store for nine years, and another who had done so for twelve. As the trade relationship develops, it becomes a personalized exchange relationship. The shop-keeper becomes more and more willing to extend credit, and for longer periods of time, always without interest;* the worker invariably does all his buying in the shopkeeper's store. The moradores know that they pay higher prices in the shops where they have established credit, but they feel obliged to continue buying there even when they have no pressing need for credit. As one said, "I feel I must help one who has helped me." But beyond this sense of moral obligation, the worker recognizes that by shopping in another store without at least trying to get the item he needs in his regular shop he is essentially breaking the *freguês* (loyal customer) bond he has established, and that credit may be denied the next time he needs it.

There are some exceptions to the above pattern. For example, three moradores told me that they habitually shop in several bodegas and receive credit in all of them. But they emphasized that they never incur a debt in one shop unless all other such debts are already paid off. Moreover, these three men expressed a clear desire not to get into debt at all, saying that they incur debts only rarely, and for very short periods and small amounts. As we saw in Chapter Six, these tiny loans are a common form of indebtedness, and do not imply a close exchange relationship between debtor and creditor. In this context it should be noted that in only one case out of 44 did I find a man owing money in two shops; he was a new arrival in extreme need, and clearly an exceptional case. A more common exception to the fregues pattern, involving a fifth of the household heads on the fazenda, is the clear refusal to incur debts in any shop. The men who do this feel that it is cheaper to shop in a bodega that does not do credit business. There are two such bodegas within easy distance of Boa Ventura; and in them, a strict business relation

---

* That is, interest, as such, is not charged. When borrowing food from other moradores, it is common to pay back in kind more than was loaned, but this kind of indebtedness is relatively uncommon. However, those bodega owners who give credit always charge higher prices—a sort of hidden interest.

exists between buyer and seller. The buyer gets the lowest available prices, the seller gets cash on the barrelhead, and that ends it.

But it would be an oversimplification to assume that lower prices are the only motive of the worker who shops in a non-credit store. An undercurrent of dislike for debt in any form runs through the economic life of Boa Ventura's workers. They distrust (one informant used the word "fear") the bondage implied by indebtedness; and they get into debt only when they have no choice. One morador put it this way: "I can't afford to buy here and there; I need credit, so I shop only in C.'s bodega." That is, there are often times when the average morador has no food at home and absolutely no cash. He has no resident landlord or company store to turn to, so he buys on credit from the local shopkeeper. Most of the time, he can then obtain cash by borrowing from the landlord on a longer term loan, selling a pig, or working for wages; usually he is able to repay the loan after a few days or weeks. Credit, then, is a practice of short-term expediency; shopkeepers are a flexible and readily available means of coping with immediate shortages of cash or food. Cash for long-term needs, investment capital, and other large sums must be found elsewhere, and about the only available source is the fazenda landlord. This will be further explored below.

To illustrate the considerations that enter into shopkeeper-debtor relations it might be helpful to give a specific example of one shopkeeper's history, attitudes, profits, and volume of business. This man, C., is a morador of Boa Ventura; and about the beginning of 1967, he decided to open up a bodega in his own house, since he lived in a section of the fazenda that was relatively populous but was nonetheless distant from all bodegas. He started out with a capital of Cr$300,000, which went fairly rapidly in loans of goods. After a few months, C. had built up a loyal clientele of about thirty persons, all of whom had bought from him on credit and hence shopped only in his store. He also had a larger group of customers who bought from him only occasionally and not on credit.

C. built his clientele so rapidly partly because he gave credit to workers who could not get credit elsewhere. He had an advantage over shopkeepers who lived off the fazenda in being a close neighbor of most of his customers. He knew the general state of their finances more closely; and in particular, he knew how much they had planted and how their crops were doing. Consequently, he was able to judge more accurately how they stood as credit risks. C. expected anyone who was in debt to him to shop exclusively in his bodega; when he heard through the very efficient gossip of the fazenda that a customer of his was asking for credit in other stores, he immediately cut that man off from further credit; he considered the customer's act a betrayal of trust, and did not want to throw good money after bad by continuing credit.

Over the period of about five months that the bodega had been operating before I interviewed C., it had been an unqualified success. Although he could not read or write names, he listed the date and quantity of each transaction in a small notebook; from this, he was able to remember who a given customer had been, and thus to remember who owed him what. He had a high turnover of goods, of which 21 per cent of the final price was profit (on the average, profits varied from none on peppers to 58 per cent on salt). He tried to set his prices more or less in accordance with prices elsewhere, but was not entirely successful at it. For instance, he did not always know how much profit he was making: as he gave me his purchase and sale prices, I figured his profits in front of him; and he was genuinely upset at the discovery that he had been charging so much for salt, which he bought in bulk and sold in very small quantities.

During the first five months, C. had excellent returns from his store, all of which he invested in 205 man-days of labor for his fields, paying wages and food for the workers out of his profits. In all, he spent roughly Cr$400,000 in cash and goods. Comparing this with an initial investment of Cr$300,000 and an additional debt of another Cr$300,000 incurred to keep his bodega stocked with food, we can see that he had a very lucrative business—if all his debtors could be trusted to pay him. At the time of my study, C. had already run out of capital and could sell no more on credit. The debts owed to him totaled Cr$600,000, exactly the amount of cash he had brought into the enterprise. Rather than investing any of his profit in creating more indebtedness among his customers, he had converted it all into labor in his own fields. His running out of capital was unfortunate, but not unexpected. In May, when the interview took place, the harvest had only just begun to come in; a long period of scarcity was barely past, no one had cash, and nearly everyone had debts. Soon men would begin to sell maize and beans, or to fatten pigs for sale, and there would be cash in circulation until the end of the cotton harvest, which might be as late as December.

Two points should be made. First, C. pumped much essential credit into the fazenda economy. To 30 regular customers he loaned an average of Cr$20,000 apiece, about the equivalent of 130 liters of maize apiece. Without some source of credit such as this, many fazenda families would risk starvation during the winter months (January to May). Second, C. himself was not the end of the credit process; in his turn, he went to a nearby town and contracted Cr$300,000 in debts in order to keep his bodega capable of offering credit. The moradores of Boa Ventura, though only in debt to members of their own small community, owed much of their credit ultimately to outside sources. In this, as in other cases, it is clear how limited is the approach, adopted for the most part in this book, of treating the fazenda as a closed unit.

THE MANAGER

By upbringing, material well-being, and personal identification, a fazenda manager is allied with, and sometimes a member of, the upper class. I interviewed six managers on three different fazendas: two had been fazendeiros in the past; one was both a manager and the owner of a neighboring fazenda; one left his position as manager to become a fazendeiro with several dependent moradores; and the other two were, by the criterion of material wealth, far above the level of moradores. As a member, or near-member, of the upper class, the manager is confident of his own superiority to the moradores, and is used to giving orders and having them obeyed. He dresses in better clothes than the workers, and rarely performs physical labor.

As a rule, a fazendeiro gives his manager free rein in running the plantation. An absentee owner may take an active interest in how the plantation is run, and he frequently determines the main lines of activity himself. But, often, he is city-bred and relatively inexpert in agricultural matters, and must take constant advice from his manager. Sometimes the manager is the only man who can get the best work out of the moradores (cf. Hutchinson 1957: 58). The workers themselves feel, somewhat cynically, that the manager's greatest pleasure is in telling others what to do, and they resent his interference with their lives.

Like the foreman of a modern industrial enterprise, a manager is caught between the standards of performance set by the owner and the standards the workers set themselves. Managers differ greatly in how much they identify with one side or the other. Some of those that I interviewed clearly felt the conflict of interests, and tried to be as easy on the moradores as they could. But many managers take a businesslike, no-nonsense attitude, insist on their own authority, and act almost entirely in the interests of the landlord. Of course, the kind of manager one finds depends very much on the kind of owner who has employed him. The workers, for their part, say that a manager is "almost an owner," and treat him with only slightly less respect than they show to the owner. They expect the manager to identify more with the interests of the landlord than with their own, but would nevertheless like to see him treat them with respect, allocate resources to them freely, and make exceptions for them when unforeseen accidents make it difficult to fulfill their obligations.

In Ceará, there are several common methods of paying a manager for his services. He may receive a fixed percentage of the harvest, giving him a personal interest in the magnitude of the crop; he may receive a fixed salary (generally not a large one); or he may be allowed to use fazenda land and labor resources free of any obligations. Commonly, some combinations of these methods is used. The manager of Boa Ventura, for instance, gains all his livelihood from the plantation. His work is purely that of directing the labor of others; for example, he decides where land is to be cleared, who is to have access to which fields, and so on. For this, he receives only a small formal wage (Cr$200,000 a year) for his and his sons' labor. However, he is allowed to use the worker's sujeição labor in his own fields when it is not needed in the landlord's plots. He pays only token wages for this labor, and the produce from his fields is not shared with the owner. Thus, although the manager does no agricultural work himself, he is one of the largest planters on the fazenda. He earns enough to keep two servants, and lives well (although not much better than that other special individual, the vaqueiro).

The manager of Boa Ventura has several responsibilities. First, he organizes and allocates the workers' sujeição. Even if a given project has been directly ordered by the landlord, the manager is always responsible for the details of its execution; usually, he accompanies the laborers to their worksite—e.g., one of the owner's fields, or a drainage ditch that must be cleaned—and oversees their labors throughout the day. A second major task of the manager is the allocation of land to the workers; he decides whether or not a worker is to have good forest to clear, or fertile bottomlands to plant. This means that he must judge whether the worker who requests a certain piece of land can exploit it efficiently; and if two workers request the same plot, he must decide how the conflict is to be resolved. The manager also allocates the fazenda housing. This activity sometimes causes great resentment among the workers, for in running the fazenda efficiently he must often move one family out of a desirable house in order to move another in.

Generally, it is the manager's responsibility to know as much as possible of everything that happens on the plantation, and he is constantly "interfering" in workers' activities when he feels they are not in the fazenda interests. It is this that earns him his reputation among the workers of "wanting to be the boss," of giving orders and making decisions simply because he enjoys exercising his power over the workers. It is clear, however, that the conflict of interests between the moradores and the fazendeiro makes constant close supervision of the moradores essential if the fazendeiro is to profit from his enterprise.

The manager of Boa Ventura was until recently a fazendeiro in his own right; but, as a result of a family dispute, he sold his land and took his present job. There is real hostility for him among the workers, which I feel derives largely from his obligation to carry out the orders of the landlord. A few moradores are extremely fond of him, and swear that he has shown great compassion at times. There were many stories told of his hardness, but a careful reexamination showed nearly all of them to be exaggerations; what such incidents usually amounted to was evicting certain moradores for reneging on their obligations.

A final matter of interest is the range of the manager's outside contacts. He is without doubt the most traveled of the fazenda residents. At least twice a month he is off to visit relatives or friends in places as far as a hundred miles away. Unlike the average morador, who remembers the names of very few kinsmen (about 25, on the average), this man can list almost every living kinsman out to his second cousins' offspring—which resulted in a list of 164 relatives. Obviously, the manager considers kin relations important and worth maintaining, in contrast to the average morador, who quickly loses contact with kin in his restless circulation from one plantation to another.*

The present landlord, Sr. Clovis Holanda, had owned Fazenda Boa Ventura for five years at the time of this study. He is a successful businessman who bought a fazenda after getting rich in commerce. He does not depend entirely on the earnings of Boa

---

* In general, members of the upper classes in Brazil have much more extensive kin ties than members of the lower classes.

Ventura for his income, for he still runs other enterprises as well; in fact, he says that he earns less from the fazenda than he would if he had invested his money in other projects. But in spite of this, he owns another fazenda elsewhere in Ceará, as well as a large sitio of fruit trees and manioc around his house just outside of Fortaleza. In addition, he owns and rents out a great many properties in Fortaleza, mainly houses and apartments. Until recently, he also had a construction business, and earned additional income as a contractor on government projects (e.g. schoolhouses). When his oldest son graduates from college as a civil engineer in the near future, he plans to reopen this business.

Clovis runs the fazenda as a profit-making enterprise. By and large, however, he invests in no capital improvements; rather, he raises cattle and crops under traditional methods, and views these products as a return on his original investment, to be invested elsewhere.

The owner's obligations to the workers. From the first day I arrived on Boa Ventura, I heard complaints from moradores about the owner of the plantation. I did not expect this, since I was by their standards a member of the upper class, and might thus be identified with the landlord to some extent. These complaints, however, were directed at Clovis, and most workers had nothing but praise for the previous owner (called "the General," since he was a general in the army). This was somewhat surprising: from all I could gather, the General had been a stern master, powerful and somewhat feared; by contrast, Clovis was more lenient, and imposed milder conditions on the workers. Why should the General be widely admired when Clovis was not? Why, in fact, should 10 out of 11 resident workers, queried separately, say that they would choose to live on a plantation run by the General rather than one run by Clovis? A solution to this paradox may be found by analyzing the workers' idea of what constitutes a "good landlord." A comparison of the General and Clovis, point by point, easily reveals the wide gap that separates the two in the eyes of the moradores.

First, a landlord should be a powerful man, holding an elite position in society. After all, one of his ideal qualities is the ability to protect his dependent workers; and the more powerful and important he is, the better he can do this. The General was a man of great power: besides holding a command in the army, he had been a police chief in Fortaleza, the capital of Ceará; moreover, he was very wealthy, and came from an old and respected family. He used this power to preserve his fazenda from outside interference. For example, one of his workers was involved in a crime, and state police came to arrest him; but the General sent them away, and even threatened to have the leader of the arresting party dismissed from his position. Many workers told me this story, and it was evident that they had been pleased to have such a master. Clovis, on the other hand, was a self-made man. Starting out as the son of a small shopkeeper, he became wealthy and bought his way into the traditional prestige system of the region by purchasing two large fazendas. The moradores seemed to feel that he was not a real landlord, since he had come from a position not much higher than theirs.

Second, a landlord must have a quality expressed by the Portuguese word *moral*. By this the moradores mean that he insists upon and receives deference, respect, and obedience, and that he expects his tenants to live upright lives (mainly, this means they should not drink alcohol, steal, or fight); if they do not live up to his expectations, he will evict them. The General was just this way. Whether it was to insist that a man marry a girl he had gotten pregnant or to evict someone who had talked back, he always thought it proper to intervene in the lives of his moradores. Clovis, by contrast, did not maintain moral: he disliked giving orders, felt uneasy in the role of an elite landowner in a relatively rigid two-class system, and actually prided himself on the degree to which he refused to interfere in the lives of the workers. In return, they showed him modest overt respect, but no deeply felt deference.

Next, a good landlord assumes responsibility for the health and well-being of his workers and their families. The General did this in several ways: by giving oranges to the sick; by using his influence in Fortaleza to get free government hospitalization and medical treatment for his moradores; and by turning milk cows over to families with babies when pasture got scarce, allowing them to feed the cows and keep the milk. The workers were extremely grateful for all these favors; that such things cost the General relatively little did not seem important to them. Clovis gave much less help of this sort, and even forbade the gardener of the fazenda orange groves to give oranges to moradores. In one case he did get a morador into the hospital, but the workers were quick to note that the General had really arranged this, even though he was no longer the landlord.

A good landlord also makes money and food available to the moradores during the very difficult period just before the harvest begins, and during the periodic drought years. The General's method was either to buy crops green, at approximately one-half of their market value when ripe, or to set up a company store in which he sold, on credit, the staples that had been given to him as shares of the last harvest. What was important to the workers was that they got cash and food when they were in danger of starving. During the drought of 1958, the General kept his moradores at work on various improvements on the fazenda; he paid low wages, but the workers survived the drought. In this respect, again, the workers did not feel they could count on Clovis. He himself had only a small cash reserve, and used it sparingly, loaning the moradores an average of less than US$11.00 apiece.* To the workers, this was not enough. After the harvest, there is much to buy, especially new clothing, and they run out of money rapidly. They count on borrowing later, and Clovis contradicts their expectations.

In the same vein, a good fazendeiro is reliable—especially when he says he will loan money to a morador, or pay him what he owes him. When the General said, "I will visit the fazenda on such-and-such a day," he arrived that day; and he paid all

* Actually, nearly one-half of this was loaned to a single entrepreneurial morador; the average loan for the others was about US$6.00.

debts promptly. Clovis often failed to arrive on the projected day. More seriously, Clovis had purchased the 1966 cotton harvest from the men in November, but by March 1967 he had still not paid them all that he owed. Expecting cash for their cotton, the workers had contracted debts in local stores in November, and were then unable to pay off those debts (which usually have a time limit on them) for several months. As a consequence, the credit of the moradores with the shopkeepers was greatly endangered. Not only had they a landlord who seemed unwilling to help them, but he had threatened their only outside source of aid, and had done so precisely at the time of their greatest need.

Finally, a good landlord is willing to spend the money and effort to make his fazenda beautiful and something to be proud of. The General had planted flowering trees, kept the roads in good shape, maintained the walls of the reservoirs, and kept fences in good repair; in short, he had given the workers a well-run fazenda, one to be proud of. And the workers had, in fact, taken pride in the greatness of their fazenda, and of their landlord. Clovis, partly because he had little ready money and partly because he did not believe in investing so much in the beautification of his property, had paid less attention to these matters during the five years of his ownership.

In short, the landlord is expected by the moradores to give more than merely rights to land. Far from hating the existence of a "company store" or the purchase of green crops at cut-rate prices, the workers regard these practices as the only alternatives to great potential suffering. Hence a landlord who does not involve them in these mechanisms, which the outside observer may be inclined to see as exploitative, is less popular. It should be obvious that the workers' view, though narrow, is quite reasonable under the circumstances. In the face of great uncertainties, a worker has no ties that can assure him as much protection as a firm tie with the landlord can. Therefore, he wants such a tie very much, even when he must pay with increased shares of his harvest and with deferential behavior.

Clovis, although he respected his workers' individuality and wanted no part of a domineering, paternalistic relationship with them, was a very frustrating man from the workers' standpoint.

Knowing how the moradores view their relationship with the landlord, we can now examine the precise nature of the "economic contract" recognized between worker and owner. The fazenda always gives the following to all moradores:

1. A house, which may be in good or poor condition, depending on chance. Houses are erected at the expense of the fazenda, but repaired at the expense of the morador. A worker is free to build his own house, but the fazenda will usually reimburse him for at least part of his labor in order to retain rights to the house.

2. A water source. This may be a well, a reservoir, or a river. In many cases, families live a great distance from their water, which must be carried to their houses in wooden kegs on the backs of donkeys. Since locations without easy access to water are unpopular, there is an abundance of good hillside land for those who will accept them.

3. Firewood, which the worker is free to cut and bring home from the fields and forests.

4. Cotton seed. Although the worker provides seed for most of his crops, it is customary for the fazenda to provide the cotton seed. Some workers select their seed from their own crop of the previous year, picking only the cleanest, healthiest seed; but most take advantage of the fazenda seed.

5. Land, by far the most important of these items. Workers have the right to farm as much land as they can reasonably expect to care for in the course of a year. The best land is scarce, however, and most of it is appropriated by the fazenda.

The worker's obligations to the owner. When the owner provides the requirements just listed, the workers assume their own obligations. On Boa Ventura, these obligations may be structured by one of two main "contracts": labor (sujeição), and sharecropping (*renda*).

There are no cases of workers who give labor only, but one-third of the moradores give two days' work a week (one-third of their total labor time) to the fazenda. In addition, these men give one-third of their manioc flour and one-half of their cotton harvest to the owner, and must sell the other half of the cotton to him at slightly below the going market rate.* Workers who operate under this contract find it onerous, but they have no choice in the matter: the manager selects them for sujeição because they live close to him and are easy to communicate orders to. On sujeição days, the fazenda provides them with their meals and Cr$500 a day, which is small reimbursement, considering that a workday in one's own plot yields an average of over Cr$5,000 in produce.

Two-thirds of the moradores work under a sharecropping contract, by which they are not required to give any days of sujeição. Instead, they give a third of their total crop of beans and corn, a third of their manioc flour, and the same cotton as they would under sujeição. Since a worker also produces potatoes, rice, lima beans, bananas, and other crops from which the fazenda requires no shares, this contract is perceived as somewhat advantageous economically. At least as important, in the eyes of the morador, is the fact that he has command over his own labor rather than being at someone's beck and call twice a week.

All of the moradores on Boa Ventura fall into one of the groups above. In addition, a worker may undertake a specific piecework task (*empleita*), to be completed at his leisure for a stated price. For example, he might agree to build a stretch of fence for Cr$3,000, working on it when his other obligations permit. Predictably, workers are very pleased to work empleita when the chance arises, which is not often.

---

* Manioc flour is not actually given to the landlord in return for the right to work land. However, the harvested manioc roots must go through a lengthy process of soaking, pounding, and milling before they can be stored as flour. The landlord provides the milling equipment, and each worker processes his own manioc. One-third of the resulting flour is given to the landlord as a fee; and since everyone on the plantation uses the mill, a third of the entire manioc harvest eventually goes to the landlord.

There are three kinds of special contracts that concern us here. First, the landlord has three fruit and vegetable gardens (sitios) on the fazenda, each run by a different morador. A sitio does not take all of a gardener's time, and each has a slightly different contract, depending on the crops grown and on whether or not he receives a wage for his work. Aside from working especially fertile, irrigated land and having free access to fruits for their own consumption, these moradores are like others on Boa Ventura—that is, their work is not especially skilled. Second, there are skilled laborers, who usually have somewhat better contracts than the unskilled moradores. Even though they must pay shares, they are always free from sujeição obligations. They receive a lower wage than their urban counterparts; but they benefit by being permitted to raise crops, working as specialists when work is available but always having agriculture to fall back on. Finally, there is the vaquiero, who is given a one-fifth share of the newborn calves. His is an excellent position: he gives neither renda nor sujeição, and is allowed to raise his calves on fazenda grazing land. Presently, he is selling his cattle as soon as possible and investing the cash in hired agricultural labor. Like the manager, he owes no shares to the landlord; hence he is like a fazendeiro in exploiting labor. The vaqueiro is the wealthiest of the moradores on Boa Ventura.

In themselves, none of these "contracts" have any long-term binding power. Nevertheless, the fazenda usually has a relatively stable agricultural population, at least during the agricultural year, and often beyond it. When a worker moves to Boa Ventura, he does not commit himself to stay any length of time. But he tends to remain because, once he starts to work his own fields, he has a stake in the harvest that he can only redeem by waiting until the harvest is gathered (unless he can find someone to "buy" his invested labor—see Chapter Four). In particular, the labor invested in establishing cotton fields will repay the worker over a period of years. Finally, most workers feel that there is a shortage of real alternatives, since most fazendas are about the same, so it is pointless to leave abruptly in the middle of a season. But in all this, the worker has no obligation to remain a stated length of time, and there is no way the landlord can force a worker to stay. This, coupled with the mild shortage of rural labor in Ceará, helps the workers of Boa Ventura to maintain an intense personal pride, which is easily wounded. Because of this pride the landlord must treat his workers with a certain amount of care and respect, even while asserting his own superior status.

In short, the workers on Boa Ventura enter into contracts with the landlord in which the rights and duties of each party are clearly delineated. A landlord who seriously neglects his duties will soon lose his workers and find replacements scarce. A landlord may or may not choose to establish protective, personalistic ties with his workers; but in any case, the relation between worker and landlord is probably best viewed as an exchange. As the workers themselves put it: "The landlord gives us a house and land, but we give him the strength of our arms. Without our arms he could do nothing."

Eight

Conclusion

The peasant sharecroppers of Boa Ventura, as we have seen, face and overcome many problems in meeting their material needs;* and in the process, they attempt to use as many resources, environmental and human, as they can. Although the workers were rarely close to starvation during my study, marginal subsistence is a fact of life to them, for the threat of drought and famine is always present. Thus the basic problem for the average sharecropper and his family is one of sheer physical survival. Under these circumstances, the workers take a very pragmatic view of their surroundings. As I have suggested, it would be a mistake to view their behavior simply as conservative and tradition-bound. In their agriculture, and in their social relations, their behavior is directed almost entirely toward meeting basic subsistence needs; they do not do things in a certain way "because things have always been done that way," but rather because this way will work reliably to assure survival.

We can see this pragmatism first of all in the workers' agricul-

---

* To the economist, needs are relative and will vary among individuals as subjective evaluations vary. This can be generalized to society, and the argument made that "needs" will be defined differently under different social systems (cf. Pearson 1957). In the case of Boa Ventura, however, "material needs" are remarkably close to the basic terms of human biology: most activity in the workers' daily life is centered around the provision of food, water, shelter, and clothing.

tural behavior. The moradores of Boa Ventura are hardworking and intelligent farmers. Their average workday is 6–8 hours, but in peak seasons it is about 10 hours. They work six days a week throughout the year, except while waiting for the rains to begin; then, some (but not all) moradores are unable to find work. Even on Sundays, when the church proscribes agricultural labor, the moradores will be found trading with one another or shopping in the bodegas.

We have seen that the tools and farming methods of Boa Ventura workers are relatively simple ones, based on hoe agriculture, and that many of the techniques in use there can be traced back to the original Indian inhabitants of Ceará. This simplicity, however, is not simply a function of the sharecropper's ignorance, for two reasons.

First, there is no easy way to determine how much improvement should be made by the workers in the existing technology (see Chapter Four). Of course, one can think of many low-cost inputs that would raise agricultural productivity without requiring much extra labor: hybrid corn seed is one example. But such inputs, in most cases, are simply not available to the workers. The moradores themselves are quite ready to try new techniques on only the vaguest hope of some benefit, as long as the experiment can be conducted at a relatively low overall cost; and their decision to accept or reject the technique is based quite sensibly on the outcome of the experiment. The real drawback is that the workers' experimentation is limited to techniques that they can afford.

Although the advanced farming methods of the temperate zones could not be transferred wholesale to any tropical region, it is certain that industrial-based farming using machinery, careful plant and animal breeding, chemical pesticides and fertilizers, etc. could raise the productivity of agriculture in Ceará. One fazenda in the region, in fact, has already begun to move rapidly in this direction, with notable success. The point I wish to make is that the decision to modernize farming is not up to the sharecropper—because of the great capital required, it is entirely beyond his control. The simple methods under which he farms are reasonable and productive methods, given the materials he has to work with.

The second major point regarding the technology of the moradores is that it is not as simple as one might think. True, a man may be a very productive farmer if he owns only a hoe and a brushhook, and can borrow an ax from time to time; but in addition, he has a complex set of agricultural decisions to make, based on a detailed knowledge of land types, crops, and weather. In general, each land type requires different varieties of crops and different planting, weeding, and harvesting techniques. Moreover, the worker must make most of his decisions at the very beginning of the agricultural cycle, for once a particular kind of land has been cleared and planted, the remainder of the worker's agricultural labor is for the most part determined.

The worker must keep several goals in mind when he allocates his land and labor. (1) He must produce an adequate food supply for the following year, which means not just enough calories, but also enough different kinds of food to offer a varied diet. (2) He must plan for an early yield of some foods during the hungry period immediately before the major harvest. (3) To be secure, he should plant in a variety of land types, but he should not waste too much time walking to and from his different fields. (4) He cannot afford to undertake either too little or too much work relative to his household labor supply. His excess labor can only be employed by others at low wages, whereas his labor shortage must be corrected by hiring others for cash; and labor and cash may be hard to find when he needs them.

These various goals are not easily met, nor is there usually a single optimum strategy. Does a man want to plant low-yielding corn for an early harvest or high-yielding corn for later? Should he reduce his crop mix to increase the productivity of one plant (e.g. manioc) or aim for a greater variety? Does he want to raise a crop that tastes better, for home consumption, or one that weighs more, for market sale? There is a wide range of choice, and a predictable variety in the actual agricultural behavior of different moradores. Making the right decisions is difficult; and there is no doubt, judging from the quantity and quality of food and clothing available in different households, that some moradores do much better than others at making these decisions.

So, although the technology of the sharecroppers of Boa Ventura appears backward when compared to that of advanced industrial society, it is nonetheless neither a poorly adapted one nor a simple-minded one. Moreover, it is not one that will be easily improved upon, for any substantial improvement would involve a huge commitment from outside the peasant sector of Brazilian society. All in all, we cannot assume that the problem is basically, or even peripherally, one of inherent peasant conservatism and lack of motivation.

But even if one accepts that the agricultural activities of the Boa Ventura sharecroppers are not examples of "peasant conservatism," there remains a large segment of behavior that appears on the face of it to be conservative—namely, the workers' tendency to orient behavior toward security. Let us briefly review the evidence that moradores seem consistently to prefer low-risk, low-yield alternatives to high-risk, high-yield alternatives.

First, we have seen that in all of the morador's agricultural activities there is one enormous uncertainty: the weather. The relationships between crops and lands vary from wet to dry years (in a very real sense the land types are not the same), and different planting strategies will be more or less productive depending on the weather. Since a worker must decide his overall strategy before he can be sure of the weather, he is faced with two broad alternatives: he can attempt to guess the weather and use a risky but potentially very profitable planting strategy; or he can plant to guard against both wet and dry growing seasons, which will always yield less than the maximum potential product but will assure some yield in nearly every year. The workers of Boa Ventura consistently pursue the low-risk strategy, for one simple reason: if they fail at the high-risk strategy, they may starve. Not merely capital but their very livelihood is at stake.

It should be noted that experimentation in crops, widely practiced on Boa Ventura, is not a risky undertaking. Experiments involve very small amounts of labor, land, and seed; and neither a failure nor a success means much during the year in which it is run. The long-term result of a repeated success, of course, is the general adoption of the new technique and an overall increase in bulk yield, resistance to disease, or some other desired end. I have also noted the workers' habit of storing a year's supply of food before selling any. This security-oriented behavior ignores the potentially more profitable alternative of selling a crop and investing the cash in labor for the next year. This happens because such an investment, which yields very well in a good year, may be a complete loss in a poor year. Other investments, e.g. livestock, offer similar possibilities for profit and similar risks. Given the profound insecurity under which he lives, the backland peasant defends himself by storing enough food for a year, and his investments in labor and livestock are usually very small.

Social relations, too, show this orientation toward security. Workers could choose to stand alone and ignore their neighbors, which would allow them the full benefit of their own labor. However, if they were to run short of food or to need a few days of labor for some urgent task, there would then be no one to help them out. Unwilling to take this risk, the moradores bind themselves in dyadic exchange relations with several other households. In the shops, workers again have alternatives. We have seen that they tend to enter into personal, enduring relationships with shopkeepers, which ensure them credit (up to a point) but cost them in higher prices. In a few stores, no credit is extended, and prices are low; but most workers do not take advantage of this situation. Finally, even with landlords the workers can distinguish two alternative kinds of relationship: one that is personalized and protective but costs more in payment of shares and general dependence, and one that is impersonal and nonprotective but costs less. Although the workers cannot always choose the kind of landlord they work for, those that I questioned consistently described the costlier, but low-risk alternative as the more desirable.

These various cases could perhaps be seen as examples of peasant conservatism, if the workers could be shown to prefer diversified planting, stockpiling, or personalism to strictly pecuniary gain for no other reason than "tradition." But it is abundantly clear that the security orientation of Boa Ventura's moradores results from the material insecurity under which they live. This insecurity has several sources. (1) The difficulty of obtaining medical treatment in case of illness constantly threatens the welfare of the workers and their families, either through sickness itself or indirectly through loss of work and income. (2) Because the moradores do not own land, they must depend on others for their basic subsistence. (3) The unpredictable weather makes agricultural forecasting impossible, and a hedging strategy is necessary in planting. (4) Crop pests and diseases attack fields from time to time. (5) Prices and money in Ceará are affected by forces that the workers do not understand, such as inflation and world-market fluctuations; therefore, they usually regard commercial and entrepreneurial ventures as far beyond their abilities.

The overall picture on Boa Ventura is one of individuals forced by their poverty and insecurity to seek, wherever possible, some protection against the many possible threats to their survival. However, in viewing the moradores of Boa Ventura we should attempt to balance their vulnerability and dependence on the one hand against their strength and independence on the other. Forced by circumstances to put security above other considerations, they nonetheless maintain a remarkable integrity and individuality. Ultimately, in the workers' outlook and practice, each household is an entirely independent unit. Nearly everything done by this unit in relation to the physical and social world beyond it is economic, and is done chiefly to increase the material well-being of the household's members. No political, religious, or social ceremonies are able to create enduring relations between any of the atomistic households of the fazenda.

One description of social relations may be phrased in terms of reciprocity. Sahlins's analysis of forms of reciprocity (1965: 47–48) isolates a "generalized reciprocity" that is found among family and friends. In this relationship, the stipulation of an equivalent return for a gift is never made explicit. Whether a generalized reciprocity exists or not may be determined by ascertaining if one-way flows are tolerated—i.e., do some individuals receive gifts without giving equivalent gifts in return?

Outside the nuclear household, exchange relations that contain one-way flows are almost nonexistent on Boa Ventura, and the exchanges are explicitly reciprocal. In this regard the treatment of kin and non-kin is identical. Anyone who does not reciprocate a gift has broken all continuing economic and social relations with the giver; any future exchanges between the two (sale, barter, wage labor, etc.) will be isolated, impersonal transactions, even if the men are brothers. The "ambulant youths" who circulate among the households of their primary kin, providing labor in exchange for food and a place to hang a hammock, fit this model: when their usefulness is over, they are generally asked to leave; similarly, they have no obligation to stay (e.g. during a harvest) and may leave whenever they want to. The tendency of families to become nuclear over time is another expression of the instrumental economic relations that predominate outside the household. Too, the minimum 100-meter distance that moradores insist on between neighbors shows the intervention of an economic concern—namely, the losses caused by a neighbors' pigs and chickens—between persons who usually have exchange relations and some social ties.

A few exceptions to the rule do occur. Gifts to the poor fit Sahlins's criteria for generalized reciprocity, although the workers do express an expectation of reciprocity from heaven at some unspecified future date. And occasionally, a beggar will return regularly to the same household, implying that his visits are part of a long-term social relationship involving one-way exchanges. A worker often charges no interest on loans made to fellow moradores, although these loans appear to be made whether or not there is an exchange (vizinho) relation between the two parties. Labor exchange is said by the moradores to benefit only one party, the one with urgent labor needs, and to be a minor inconvenience to the other; but it is encountered with even greater frequency than the hiring of wage laborers. There may also be other one-way flows that I overlooked. But it is clear that the one-way flows are relatively uncommon, and that "equal exchange" is the conventional mode of social interaction on Boa Ventura.

The expectation of equal return applies to vertical economic relations as well. The workers must depend on the shopkeepers and the landlord in their daily life, and they form definite ties with these patrons; but they feel that in return they are providing a reliable clientele and labor force, as well as higher profits for their superiors. The moradores cannot, of course, demand everything they wish from their patrons, but they can and do demand respect and fair play. During the year of my study, there was a mild labor shortage in the region near Boa Ventura, which allowed a dissatisfied tenant to leave one fazenda and find work on another quite easily. This power to migrate usually exists, and it maintains the morador's sense of pride and independence. The workers expect to be treated courteously and fairly by the landlord, and will leave if he offends them. A worker is never condemned when he attempts to improve his living conditions by moving to a place where he thinks things will be better.

The desire for security that draws a worker into exchange relations with neighbors, storekeepers, and the landlord often conflicts seriously with his desire for independence, which he tries to maintain through his right to break off horizontal and vertical ties at any time. The workers express a particular distaste for indebtedness, which they consider a kind of slavery; and they place a high value on "escaping" the fazenda system and becoming independent landowning farmers. Nevertheless, the low productivity of small landholdings and the general environment of insecurity in Ceará force them to seek protection. This conflict between two contradictory goals gives the workers' total behavior an ambivalent aspect, which they themselves feel perhaps more keenly than anyone.

Glossary
Bibliography
Index

Glossary

Baixo. Moist, low-lying land
Bodega. A small shop specializing in necessities like salt or matches
Broca fina. Light clearing of hillside land, involving weeds and scrub
Broca grossa. Heavy clearing work—trees, stumps, etc.
Caatinga. The thorny scrub-brush cover characteristic of the sertão
Campestre. Worthless hillside land
Campo. A plot with no trees or stumps, suitable for plowing
Capoeira. A cleared hillside plot, second year and after
Coroa. Moist lands on the margins of rivers
Cruzeiro. Brazilian monetary unit, exchanging in 1967 at Cr$2,700
    to the U.S. dollar
Dono de casa. Head of the household
Empleita. Piecework
Fazenda. A large landholding
Fazendeiro. The owner of a fazenda
Freguês. Loyal customer
Lagoa. Land covered with water during part of the year
Lastro. A field planted in only one crop
Manga. A large fenced pasture
Mata. Forest
Minifúndia. Small independent landholdings.
Morador. Resident worker on a fazenda
Moral. Morals, upright behavior
Patrão. Patron (for a fazenda worker, usually the landlord)
Renda. Share of a crop paid to landlord in return for rights to farm-
    land
Rio. Farmland in the empty riverbed, planted only during the dry
    season

Roçado. A hillside swidden plot during the first year after clearing

Sertão. The semi-arid backlands of northeastern Brazil

Sitio. An irrigated fruit garden

Sujeição. Labor obligation assumed in return for rights to farmland

Troca. Barter

Vaqueiro. Cowboy

Vazante. Land on the margin of a reservoir

Vizinho. Neighbor

Bibliography

Adams, Richard N. 1964. Rural labor. *In* John J. Johnson, ed., Continuity and change in Latin America. Stanford, Stanford University Press.

Andrade, F. Alves de. 1954. A propriedade rural no polígono das sêcas. Fortaleza, Escola de Agronomia do Ceará.

———— 1960. Agropecuária e desenvolvimento do nordeste. Fortaleza, Imprensa Universitária do Ceará.

Andrade, Manoel Correia de. 1963. A terra e o homem no nordeste. São Paulo, Editôra Brasiliense.

Baer, Werner. 1965. Industry and economic development in Brazil. Homewood, Ill., Irwin.

Barth, Frederik. 1956. Ecological relationships of ethnic groups in Swat, North Pakistan. *American Anthropologist* 58: 1079–89.

Belshaw, Cyril. 1965. Traditional exchange and modern markets. Englewood Cliffs, N.J., Prentice Hall.

Berreman, G. 1966. Anemic and emetic analysis in social anthropology. *American Anthropologist* 68: 346–54.

Beshers, James M. 1967. Population processes in social systems. New York, The Free Press.

Boserup, E. 1965. The conditions of agricultural growth. Chicago, Aldine.

Burling, Robbins. 1962. Maximization theories and the study of economic anthropology. *American Anthropologist* 64: 802–21.

Cancian, Frank. 1965. Economics and prestige in a Maya community. Stanford, Stanford University Press.

Carneiro, Robert. 1961. Slash and burn cultivation among the Kuikuru and its implications for cultural development in the Amazon Basin. *In* Johannes Wilbert, ed., The evolution of horticultural systems in native South America. Caracas, Sociedad de Ciencas Naturales La Salle.

CIDA. 1966. Posse e uso da terra e desenvolvimento socio-econômico do setor agrícola: Brasil. Comitê Interamericano de Desenvolvimento Agrícola. Washington, D.C., Pan American Union.

Conklin, Harold C. 1957. Hanunóo agriculture in the Philippines. FAO Forestry Development Paper 12. New York, United Nations.

Da Cunha, Euclides. 1944. Rebellion in the backlands. Chicago, University of Chicago Press.

Diegues, Manuel, Jr. 1959. Land tenure and use in the Brazilian plantation system. *In* Plantation systems of the New World. Pan American Union Social Science Monograph 7.

Erasmus, Charles. 1965. The occurrence and disappearance of reciprocal farm labor in Latin America. *In* Dwight Heath and Richard Adams, eds., Contemporary cultures of Latin America. New York, Random House.

Firth, Raymond, and B. S. Yamey, eds. 1964. Capital, savings, and credit in peasant societies. Chicago, Aldine.

Fitchen, Janet M. 1961. Peasantry as a social type. *Proceedings of the Annual Spring Meeting of the American Ethnological Society*, pp. 114–19. Seattle, University of Washington Press.

Foster, George M. 1961. The dyadic contract: A model for the social structure of a Mexican peasant village. *American Anthropologist* 63: 1173–92.

——— 1963. The dyadic contract in Tzintzuntzan, II: Patron-client relationship. *American Anthropologist* 65: 1280–94.

——— 1965. Peasant society and the image of limited good. *American Anthropologist* 67: 292–315.

Frake, Charles. 1962. Cultural ecology and ethnography. *American Anthropologist* 64: 53.

Fried, J. 1962. Social organization and personal security in a Peruvian hacienda Indian community: Vicos. *American Anthropologist* 64: 771–80.

Geertz, Clifford. 1963. Agricultural involution. Berkeley, University of California Press.

Georgescu-Roegen, Nicholas. 1964. Economic history and agrarian economics. *In* Carl Eicher and Lawrence Witt, eds., Agriculture in economic development. New York, McGraw-Hill. First published in 1960.

Gillin, John. 1951. The culture of security in San Carlos. Publication No. 16 of the Middle American Research Institute. New Orleans, Tulane University

Girão, Raimundo. 1947. História econômica do Ceará. Fortaleza, Instituto do Ceará.

Goldschmidt, Walter. 1965. Theory and strategy in the study of cultural adaptability. *American Anthropologist* 67: 402–8.

Harris, Marvin. 1956. Town and country in Brazil. New York, Columbia University Press.

———— 1964. Patterns of race in the Americas. New York, Walker.

———— 1968. The rise of anthropological theory. New York, Crowell.

Homans, George C. 1958. Social behavior as exchange. *American Journal of Sociology* 63: 597–606.

Hutchinson, Harry. 1957. Village and plantation life in Northeast Brazil. Seattle, University of Washington Press.

IPE. 1964. Diagnóstico socio-econômico do Ceará. Instituto de Pesquisas Econômicas da Universidade do Ceará. Fortaleza, Imprensa Universitária do Ceará.

Johnson, Allen W., and Bernard J. Siegel. 1968. Wages and income in Ceará, Brazil. *Southwestern Journal of Anthropology* 25: 1–13.

Jones, William O. 1959. Manioc in Africa. Stanford, Stanford University Press.

Le Clair, E., Jr. 1962. Economic theory and economic anthropology. *American Anthropologist* 64: 1179–1203.

Leeds, Anthony. 1961. Introduction. *In* Johannes Wilbert, ed., The evolution of horticultural systems in native South America: Causes and consequences—a symposium. Caracas, Sociedad de Ciencias Naturales La Salle.

Leeds, Anthony, and Andrew P. Vayda, eds. 1965. Man, culture, and animals: The role of animals in human ecological adjustments. Washington, D.C., American Association for the Advancement of Science.

Lewis, Oscar. 1951. Life in a Mexican village: Tepoztlán restudied. Urbana, University of Illinois Press.

Manners, R. 1956. Tabara: Subcultures of a tobacco and mixed crops municipality. *In* Julian Steward, *et al.*, The people of Puerto Rico. Urbana, University of Illinois Press.

Mauss, Marcel. 1954. The gift. Glencoe, Free Press.

Meggers, Betty. 1957. Environment and culture in the Amazon Basin: An appraisal of the theory of environmental determinism. *In* Studies in Human Ecology. Pan American Union, Social Science Monograph 3. Washington, D.C.

Metraux, Alfred. 1948a. The Teremembé. *In* Julian Steward, ed., Handbook of South American Indians. Bureau of American Ethnology, Bulletin No. 143.

———— 1948b. The Tupinambá. *In* Julian Steward, ed., Handbook of South American Indians. Bureau of American Ethnology, Bulletin No. 143.

Miller, Solomon. 1967. Hacienda to plantation in northern Peru: The processes of proletarianization of a tenant farmer society. *In* Julian Steward, ed., Contemporary change in traditional societies, Vol. 3. Urbana, University of Illinois Press.

Mintz, Sidney W. 1953. The culture history of a Puerto Rican sugar plantation, 1876–1949. *Hispanic American Historical Review* 33.

Nicholls, William H., and Ruy Miller Paiva. 1966. Ninety-nine fazendas: The structure and productivity of Brazilian agriculture, 1963. Vanderbilt University, Graduate Center for Latin American Studies.

Oberg, Kalervo. 1965. The marginal peasant in rural Brazil. *American Anthropologist* 67: 1417–27.

PAUSSM. 1959. Plantation systems of the New World. Pan American Union, Social Science Monograph 7. Washington, D.C.

Pearson, Harry. 1957. The economy has no surplus. *In* Karl Polanyi, Harry Pearson, and Conrad Arensberg, eds., Trade and markets in the early empires. Glencoe, Free Press.

Pospisil, Leopold. 1963. Kapauku Papuan economy. Yale University Publications in Anthropology. New Haven, Yale University Press.

Prado Junior, Caio. 1963. História econômica do Brasil. São Paulo, Editôra Brasiliense.

Rappaport, Roy. 1967. Pigs for the ancestors: Ritual in the ecology of a New Guinea people. New Haven, Yale University Press.

Robock, Stefan H. 1963. Brazil's developing Northeast. Washington, D.C.: Brookings Institution.

Rogers, Everett. 1969. Modernization among peasants: The impact of communication. New York, Holt, Rinehart, and Winston.

Sahlins, Marshall. 1962. Moala. Ann Arbor, the University of Michigan Press.

———— 1965. On the sociology of primitive exchange. *In* The Relevance of models for social anthropology (A.S.A. Monograph 1). New York, Praeger.

Salisbury, R. 1962. From stone to steel. New York, Melbourne University Press.

Schultz, Theodore W. 1964. Transforming traditional agriculture. New Haven, Yale University Press.

Simonsen, Roberto C. 1962. História econômica do Brasil, 1500–1820. São Paulo, Companhia Editôria Nacional.

Steward, Julian H. 1955. Theory of culture change. Urbana, University of Illinois Press.

Steward, Julian H., *et al.* 1956. The people of Puerto Rico. Urbana, University of Illinois Press.

Tax, Sol. 1953. Penny capitalism. Institute of Social Anthropology Publication 16. Washington, Smithsonian Institution.

Thomlinson, Ralph. 1965. Population dynamics. New York, Random House.

Wagley, Charles. 1957. Plantation America: A cultural sphere. *In* Vera Rubin, ed., Caribbean studies. Jamaica: Institute of Social and Economic Research, University College of the West Indies.

Wolf, Eric R. 1955. Types of Latin American peasantry. *American Anthropologist* 57: 452–71.

——— 1956. San Jose: subcultures of a "traditional" coffee municipality. *In* Julian H. Steward, ed., The People of Puerto Rico. Urbana, University of Illinois Press.

——— 1966. Peasants. Englewood Cliffs, N.J., Prentice-Hall.

Wolf, Eric R., and Sidney Mintz. 1957. Haciendas and plantations. *Social and Economic Studies* 6: 380–412.

Index